Pro/ENGINEER® Tips and Techniques

Tim McLellan and Fred Karam

Pro/ENGINEER® Tips and Techniques

Tim McLellan and Fred Karam

Published by:

OnWord Press
2530 Camino Entrada
Santa Fe, NM 87505-4835 USA

First Edition, 1996
SAN 694-0269
10 9 8 7 6 5 4 3 2

Printed in the United States of America

Library of Congress Cataloging-in-Publication Data

McLellan, Tim, 1966-
Pro/Engineer tips and techniques / Tim McLellan and Fred Karam.
p. cm.
Includes index.
ISBN 1-56690-053-0 (alk. paper)
1. Pro/Engineer. 2. Computer-aided design. 3. Mechanical drawing--Computer programs. I. Karam, Fred, 1965- . II. Title.
TA174.M3836 1996
620'.0042'028553--dc20

95-51494
CIP

Trademarks

Warning and Disclaimer

About the Authors

Tim McLellan is an engineering designer for Ford Motor Company and a former senior application engineer for Parametric Technology Corporation. He has a B.S. in industrial engineering technology with a minor in computer science from Central Michigan University. Tim's experience includes a diverse product design background as well as procedure, training, and implementation development.

Fred Karam is president of Engineering Solid Solutions, Inc., a Detroit-based engineering consulting firm that specializes in CAD/CAM/MCAE. He holds a B.S. degree in mechanical engineering from Lawrence Technological

University, and will soon complete an M.S. in manufacturing systems engineering at the University of Michigan. Prior to founding ESS, Fred worked for several companies in the areas of automotive power train design, Class I body surfacing, castings, stampings, plastic injection molding, and rapid prototyping.

Acknowledgments

I would like to thank my loving wife, Deb, for her support and patience during this project. I would also like to thank my manager and project leaders, Steve, Chuck, and Dean, at Ford Motor Company. They were especially understanding when, due to time constraints, I would leave at 2:30 to write a chapter. Finally, I extend a special thanks to Craig Widmaier and the rest of Parametric Technology Corporation's Detroit office.

Tim McLellan

First, I would like to express my appreciation to Tim McLellan for giving me the opportunity to participate in writing this book. Next, I would like to say a big "thank you" to everyone at work and at home who has supported me over the past year, especially my wife Christine.

Fred Karam

OnWord Press Credits

Dan Raker, President
Gary Lange, Associate Publisher
David Talbott, Acquisitions Editor
Barbara Kohl, Project Editor
Carol Leyba, Production Manager
Michelle Mann, Production Editor
Janet Leigh Dick, Marketing Director
Lynne Egensteiner, Cover designer, Illustrator
Kate Bemis, Indexer

Table of Contents

Introduction xiii

Part 1: Augmenting the Development Baseline

Chapter 1:
The Design Blueprint: Capturing Your Design Intent 3

Introduction 3
Design Breakdown 4
Avoid the Cut and Paste Method 5
Keep Features Simple 6
Developing Standards 7
Standards Directory 7
Standards for config.pro Files 8
Standards for menu_def.pro Files 9
Other Standards 9
Summary 9

Chapter 2:
Going Beyond the Basic Sketcher 11

Introduction 11
Building Flexibility into a Sketch 11
Construction Tools 12
Information Tools 13
Other Tools 14
When Sketches Do Not Regenerate 17
The Sketcher Flow 18
Summary 19

Part 2: Product Development—Part Mode

Chapter 3:
Following a Blueprint into Part Mode 23

Introduction 23
Setup and Concurrent Engineering 23

Flexing the Part . . . 27
Summary . . . 28

Chapter 4:
Complete Part Creation Using the Shell Function . . . 29

Introduction . . . 29
Planning Shelled Parts . . . 31
Creating a Stamped Metal Part . . . 32
Changing the Shelled Part After Testing . . . 35
Changing the Shelled Metal Cover: Adding a Clearance Hole . . . 41
Summary . . . 43

Chapter 5:
Working with Non-Planar Shapes . . . 45

Introduction . . . 45
Development Core: The Blueprint . . . 46
Developing a Laundry Detergent Bottle . . . 47
Accommodating the Design Changes . . . 51
Major Change in Design . . . 55
Additional Control Through Graph Features and Relations . . . 58
Adding A Handle . . . 63
Rugged Appearance . . . 64
Final Details . . . 67
Summary . . . 69

Chapter 6:
Feature Reduction When Creating Complex Shapes . . . 71

Introduction . . . 71
Fan Design Requirements . . . 72
Fan Construction . . . 73
Hub Design . . . 74
Blade Design . . . 76
    Construction of the First Blade . . . 76
    Adding More Blades . . . 81
Other Feature and Regeneration Time Reduction Techniques . . . 82
Summary . . . 84

Chapter 7:
Developing Design Options Using Pro/PROGRAM . . . 85

Introduction . . . 85

Misconceptions about Pro/PROGRAM 86
Controlling a Cube Using Pro/PROGRAM. 87
Expanding the Use of Pro/PROGRAM 91
The Golf Club Blueprint 91
Using Pro/PROGRAM to Resolve Design Requirements 92
Is the Golf Club a Wood? 92
Golf Club Loft or Number. 96
Deciding on Cavity Backs for Irons. 98
Adding Features to a Pro/PROGRAM 100
Summary. 102

Chapter 8:
Patterning Complex Sweeps 103

Introduction 103
Sweep Blueprint 104
Working Through Design Requirements. 105
3D Center Line. 106
Splitting the 3D Center Line into Portions 107
Creating the Bump to be Patterned. 109
Relating Design Variables for a Successful Pattern 112
Summary. 116

Chapter 9:
Creating a Bending Spring 117

Introduction 117
Blueprint for the Bending Spring 118
Setting up the Bending Spring 119
Developing the Foundation for the Bending Spring Shape 122
Using Setup and Foundation to Generate the Bending Spring. 123
Creating the Solid Portion of the Bending Spring 127
Summary. 129

Part 3: Free-Form Design

Chapter 10:
Advanced Curves and Surfaces 131

Introduction 133
Curves. 134
Conic Sections 134
Splines. 137

Curve Analysis . . . 142
Curvature Display . . . 143
Conic Curvature Accelerations . . . 144
Spline Curvature Accelerations . . . 145
Example of Straight Line and Arc . . . 146
Example of Multiple Curves . . . 148
Surfaces . . . 151
Curves . . . 152
Control Points . . . 153
Tangency . . . 154
Tips for Creating Boundary Blended Surfaces . . . 154
Surface Analysis . . . 155
Max Dihedral . . . 155
Porcupine Smoothness . . . 156
Color Rendering Analysis . . . 157
Reflect Curves . . . 158
Summary . . . 159

Chapter 11: Starting a Surface Design . . . 161

Introduction . . . 161
Setup . . . 162
Default Datums . . . 162
Goals for Working with Complex Shapes . . . 163
Framing . . . 165
Example of Creating A Blended Surface . . . 168
Summary . . . 175

Chapter 12: The Surface Design Project . . . 177

Introduction . . . 177
Creating a Fully Parametric Base Cap Surface . . . 177
Leg Development . . . 179
Web Development . . . 189
Ring Development . . . 193
Summary . . . 196

Part 4: Assembly Functions

Chapter 13:
Following A Blueprint Into Assembly Mode 199

Introduction 199
Concurrent Engineering for Setup and Assembly 200
Flexing the Assembly. 203
Summary. 205

Chapter 14:
Kinematics: Making Assemblies Move 207

Introduction 207
Approaching the Moving Assemblies Design Problem 208
How the Skeleton Technique Works 209
Creating the Kinematic Assembly. 212
Creating the Kinematics in the Assembly 212
Benefits of the Skeleton Part 213
Expanding the Skeleton Part into 3D 214
A Kinematic Short Long Arm (SLA) Automotive Suspension Assembly 214
Initial Skeleton Design Position. 215
Long Arm Skeleton Development 216
Short Arm Skeleton Development. 218
Adding the Knuckle (Fixed-length Component) 220
Verifying and Testing the Kinematics Assembly 221
Automating the Kinematics Assembly. 223
Summary. 224

Chapter 15:
Working with Large Assemblies. 225

Introduction 225
The Advanced User's Basic Guidelines. 226
Advanced Large Assembly Management Tools 227
Using Simplified Representation (Include | Exclude). 227
Expanding Simplified Representations Using Rules. 230
Using By Rule - Model Name 230
Using Rule - Size 231
Using By Rule - Distance 233
Using By Rule - Expression. 234
Using the By Rule Zone Option 235

Additional Options in Simplified Representation 237
Creating an Envelope 237
Using the Envelope Part as a Substitute 239
Using Other Representations in Simplified Representations 240
Summary 241

Chapter 16:
Master Modeling 243

Introduction 243
Process Overview 244
Description of Master Modeling 245
Tips and Techniques 246
Telephone Assembly Example 246
Summary 249

Part 5: Pro/NOTEBOOK

Chapter 17:
Using Layout Mode to Evaluate Designs 253

Introduction 253
Why Layout? 254
Creating a Layout 255
Layout Goals 256
Achieving Layout Goals 256
2D Illustration of Components and Assembly Mounting References 256
Design Guidelines 259
Charting Available Detergent Caps 259
Table Input Section 260
Design Checklist 262
Cost Evaluation Chart 264
Objectives of Completed Layout 266
Using Layout Global Information in Components 268
Referencing a Component to the Layout (Declaring) 268
Obtaining Assembly Mounting Information for Automatic Assembly 268
Selecting the Detergent Cap Used in the Final Assembly 270
Obtaining Critical Parameters for "Height" and "Width" 274
Summary 277

Part 6: Information Tools

Chapter 18:
Information Tools 281

Introduction 281
Regeneration Information (Regen Info) 281
Parent/Child Information 283
Information via Resolve Feature Mode 285
Information When a Failure Occurs 285
Working with Resolve Feature Mode 285
Additional Information in Resolve Feature Mode 287
Summary 295

Part 7: Interface Options

Chapter 19:
Working with Pro/MESH 299

Introduction 299
Designing for FEM 300
Model Clarity 300
Model Content 305
Meshing Tips 306
Summary 307

Index 309

Introduction

Pro/ENGINEER Tips and Techniques provides the advanced user with a valuable set of real world scenarios that can be used to improve any product development process. The tips and techniques illustrated throughout the book elevate the advanced user to new heights using Pro/ENGINEER. The scope of the book provides methods applicable to many situations in an easy to understand fashion. Virtually every facet of Pro/ENGINEER is examined, from advanced techniques using the Sketcher to free-form surface development.

In every chapter a scenario is developed around a specific product development or design issue. Advanced users can then apply the tips and techniques demonstrated to similar situations. The scenarios developed throughout the book are based on years of experience in engineering and reflect common problems that advanced Pro/ENGINEER users wish to master. Although the examples may not provide every menu selection needed to complete a task, advanced users obtain the foundation on which to expand knowledge and improve techniques.

Book Structure

This book is divided into seven parts. Part 1 focuses on the development baseline with tips on the blueprint, design intent, and going beyond the basic sketcher. Part 2 addresses product development with discussion of following a blueprint into part mode, part creation using the shell function, working with non-planar shapes, feature reduction when creating complex shapes, developing design options using Pro/PROGRAM, patterning complex shapes, and creating a bending spring.

Part 3 on free-form design provides tips and techniques on advanced curves and surfaces, starting a surface design, and surface design projects. Part 4 is dedicated to assembly functions, including following a blueprint into assembly mode, kinematics, working with large assemblies, and master modeling.

Part 5 is focused on Pro/NOTEBOOK. Part 6 addresses information tools, and Part 7, interface options.

Finally, the book concludes with a detailed index.

Typographical Conventions

The names of Pro/ENGINEER menus, menu options, functions, modes, commands, and so on are capitalized. Examples are bulleted.

- ❑ In the Group menu, select the Pattern option for groups.
- ❑ To carry out the design changes, take advantage of options such as Insert mode, the Graph feature, and Redefine.
- ❑ The Align, Use Edge, and Offset Edge functions can also help to build flexibility into a sketch.

Short command sequences (menu selections) are linked with a small arrow in regular text. Long command sequences appear on a separate line. Examples appear below.

- ❑ Finally, select **Done** → **Return**.

Set Up → Parameters → Part → Create → Number

User input and names for files, parameters, directories, variables, and so on are italicized. Examples appear below.

- ❑ Use the *wall_thk* part parameter when inputting the following value: *(10 + (2*wall_thk))*.
- ❑ When prompted for the value of *wall_thk*, key in *3*.
- ❑ After entering a name (*BUMPS*), Pro/ENGINEER will activate the sketcher.
- ❑ The *loadpoint/text/config.sup* file contains options that cannot be overridden.

Emphasis is indicated by italics. Examples are bulleted.

- *Dimensioning* is the most widely used function in the Sketcher to accomplish flexibility.
- Setting up a part while considering all aspects of product development can help users achieve *concurrent engineering*.
- Wireframe data should *always* be read as datum curves from files.

In some cases, computer prompts and responses as well as user input in a Pro/PROGRAM appear in a monospaced font. In the following example, computer program code is flush left, and user responses are indented and boldfaced.

```
VERSION XX.0
REVNUM XX
LISTING FOR PART CUBE
INPUT
END INPUT
RELATIONS
        W = H
        D = H
END RELATIONS
```

✗ **TIP:** *Tips on functionality usage, shortcuts, and other information aimed at saving you time appear like this.*

✔ **NOTE:** *Information on features and tasks that is not immediately obvious or intuitive appears in notes.*

! **WARNING:** *A handful of warnings appear in this book. They are intended to help you avoid committing yourself to undesirable results.*

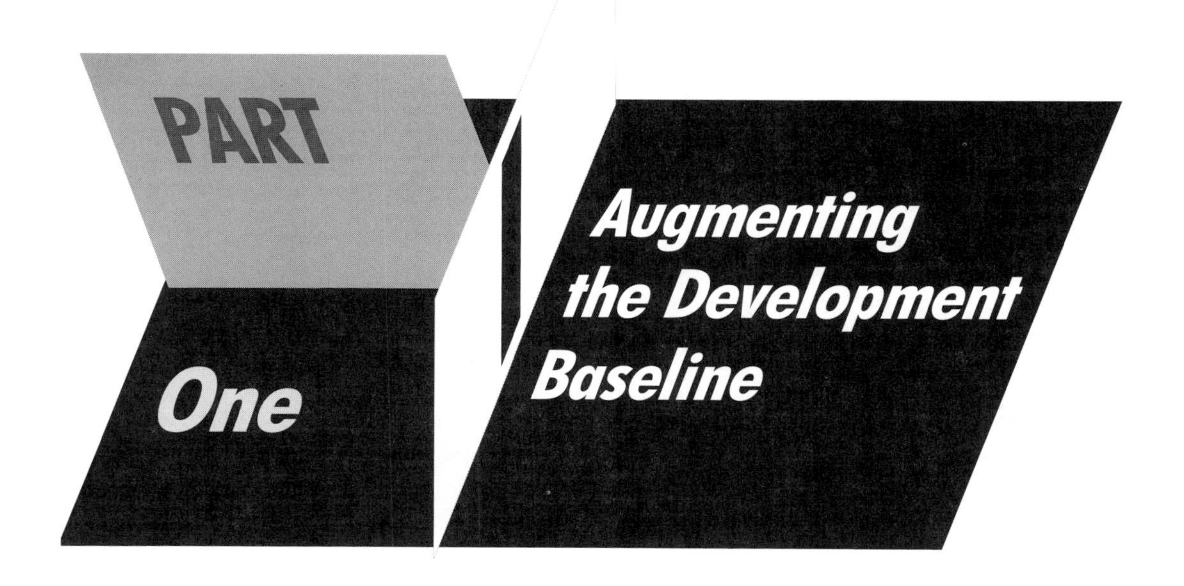
PART
One
Augmenting
the Development
Baseline

The Design Blueprint: Capturing Your Design Intent

Introduction

Capturing all production, assembly, and manufacturing requirements for product design can be described as a blueprint. The blueprint serves as a guideline to capture the design intent, and the degree of design success is correlated with following the blueprint before, during, and after product development. Easy revisions to parts and assemblies are examples of design success made possible by following the blueprint. Basic procedures in developing a successful design appear below.

1. Break down the design.
2. Avoid the cut and paste method.
3. Keep features simple.
4. Develop standards.

We will discuss sample problems throughout this book. Underlying each problem is a blueprint for product development.

Design Breakdown

The first and most important step in a product development blueprint is design breakdown. Users are taught to develop a part by considering the largest features first, and then moving on to smaller features. An analogy for this process could be building a puzzle. When creating a puzzle, you start with a single piece (the most important one), and then use this piece to guide the rest of the development. Pro/ENGINEER can be used in a similar fashion.

Identify the most important feature of the design, which may not be the largest feature. Relating or developing parent/child relationships from that feature will help to ensure flexibility. Next, identify the second most important feature and so on. Advanced users will often sketch the design breakdown portion of the blueprint on paper or make mental notes based on total design requirements.

✔ **NOTE:** *While breaking down the design, make sure that you weigh in all parts of product development including assembly and manufacturing. This procedure will increase the overall flexibility of the design. For instance, if you focus exclusively on building the geometry for the marketing department, changes required by manufacturing may become difficult.*

A design blueprint example is to create two bosses on top of a block so that if one boss is moved the other moves with it. An easy way of accomplishing this would be to create one boss and then dimension the second boss to the first.

A design which accomplishes the blueprint.

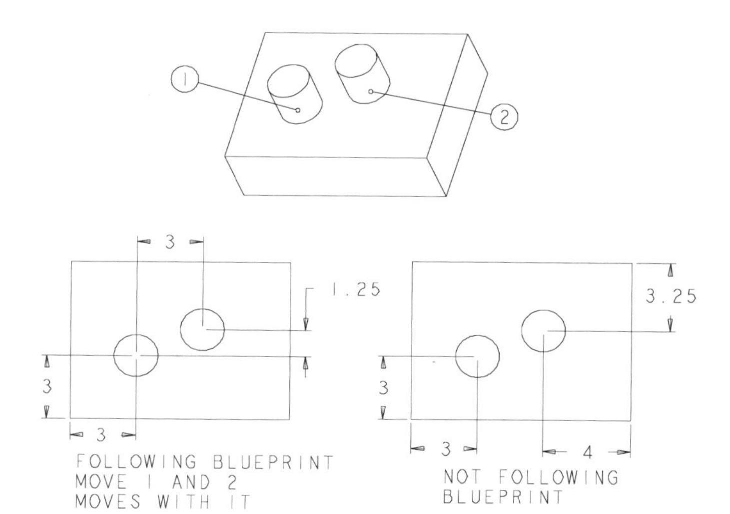

Avoid the Cut and Paste Method

Avoiding the cut and paste method is almost as important as breaking down the design. The cut and paste method can best be described as building a part while ignoring the blueprint. A good metaphor for cutting and pasting relates to manufacturing. A welder would not weld material to a part and then cut it off. A designer, in turn, would not add a protrusion and then cut it off. Advanced users find that building or working on a part which employed the cut and paste method causes problems during product revision.

To avoid the cut and paste method, you must consider each feature's role in the blueprint. Does the feature have a purpose in product development? Does the feature contribute to geometry creation, assembly, and/or manufacturing? Your goal is to answer yes to each of these questions.

Each feature created during product development should have a purpose and a desired action/reaction. Each feature's role refers to future features. Your objective is for subsequent features to react appropriately with earlier features.

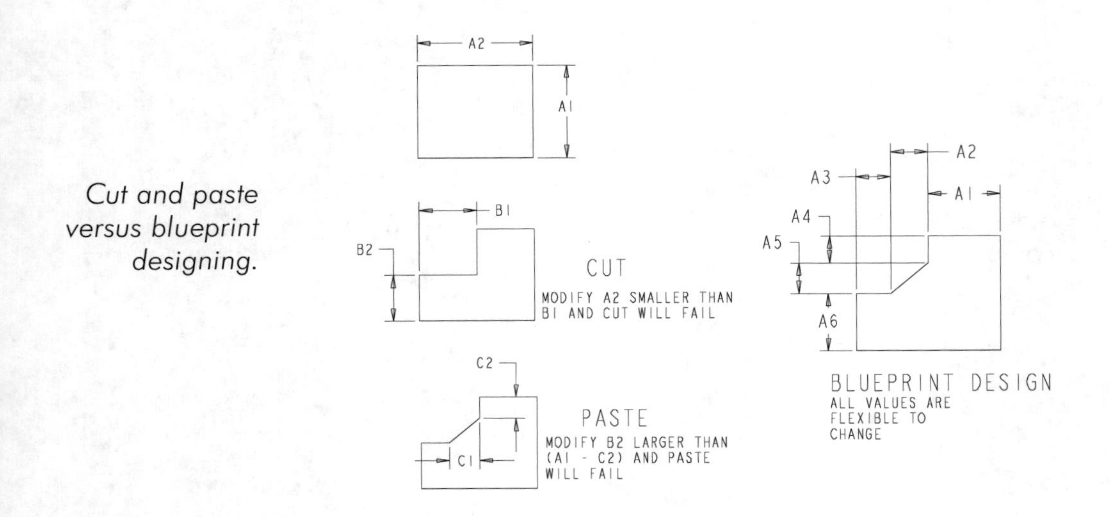

Cut and paste versus blueprint designing.

Keep Features Simple

Keeping features simple does not mean that the design is simple. An important rule of thumb is to place one important piece at a time. Adding two simple features rather than a single complex feature is often more effective in terms of design flexibility and ease of making changes.

Keeping features simple allows for easier modifications. In addition, if more than one person eventually works on the design, feature simplicity facilitates understanding of the original creator's design.

Complex cross section versus two simple features.

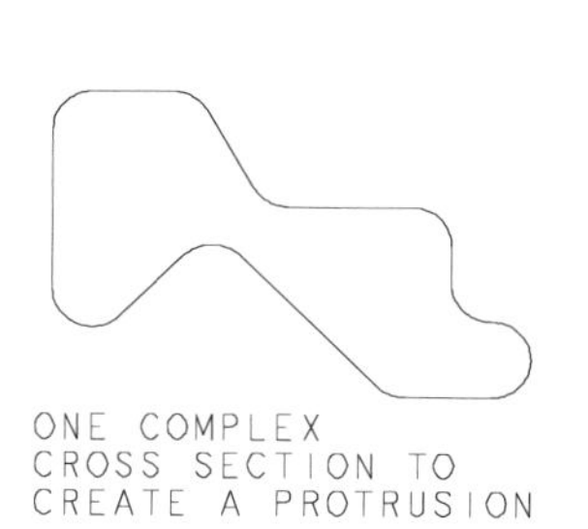

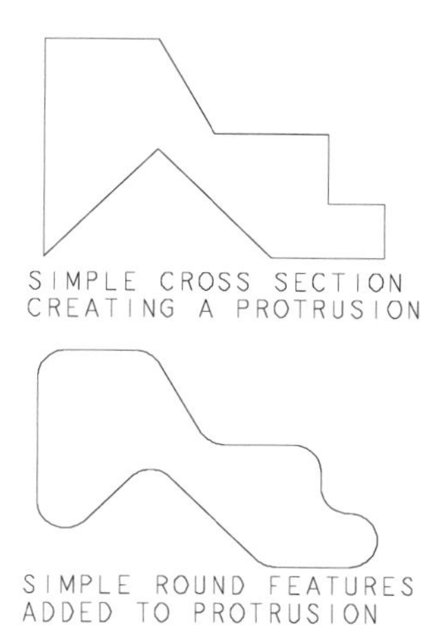

SIMPLE CROSS SECTION
CREATING A PROTRUSION
SIMPLE ROUND FEATURES
ADDED TO PROTRUSION

Developing Standards

Although often overlooked by users, standards play an important role in the product development blueprint because they can help make tasks easier and improve overall productivity. Examples of the most useful standards are discussed below.

Standards Directory

The standards directory—which should be in place on every machine or a file server—includes diverse subdirectories containing company-specific information. The subdirectories can be automatically searched through *config.pro* options and allow for quick and easy access. To access the standards directory, use the following command string:

```
server/pro_standards/default_start_data
```

✔ **NOTE:** *The default_start_data directory may contain default parts, assemblies, sheet metal, and manufacturing parts that are already set up with datum planes and named views. These parts can be set up for retrieval through a Pro/ENGINEER menu pick.*

Subdirectories containing formats, udfs, and material standards are listed below. The last subdirectory of the four would contain "other" company-specific information.

```
server/pro_standards/formats
server/pro_standards/udfs
server/pro_standards/material
server/pro_standards/*
```

Standards for config.pro Files

The four basic *config.pro* files are read in a specific order and are located in specific paths.

- ❑ The *loadpoint/text/config.sup* file contains options that cannot be overridden.

✔ **NOTE:** *Contents of the config.sup file should be kept to a minimum. Most companies choose not to use this option.*

- ❑ The *loadpoint/text/config.pro* file should contain global mapkeys and all the global options that your company desires, such as *search_path*, *pro_format_dir*, and *pro_material_dir*.

✘ **TIP:** *Search paths established in the config.pro file should be arranged from the most to the least frequently used. This technique improves retrieval time.*

- ❑ The *home directory/config.pro* file contains personal settings.
- ❑ The Pro/ENGINEER *startup directory/config.pro* file contains settings specific to start-up or group settings.

Consider adding layers to *config.pro* files in order to automatically place various types of items on a layer. Next, such layers can also be sublayered, thereby automatically creating a nesting effect and saving you a lot of time during product creation.

See Appendix D, "Configuration File Options," in the *Pro/ENGINEER Fundamentals* user's guide for complete references.

Standards for menu_def.pro Files

Settings in the *menu_def.pro* files can play an important part in standardizing or customizing your environment. For instance, to make an option or series of options available from a Pro/ENGINEER menu pick, use *menu_def.pro* options.

- ❑ The *loadpoint/text/menu_def.pro* file should contain global options desired by your company, such as a menu pick to automatically shade the object.
- ❑ The *home directory/menu_def.pro* file contains personal *menu_def* settings.

Consider a menu pick addition which automatically retrieves a default part and the user to rename it for use as a new part. The default part would contain default datum planes and stored views. Command lines for setting this menu pick addition follow:

```
setbutton MODE New_Part "#MODE;#PART;\
#RETRIEVE;server/pro_standards/default_start_data/default_part;\
#DBMS;#SAVE AS;;#ERASE;#CONFIRM;#PART;#SEARCH|RETR;"\
"Name and retrieve new default starting part."
```

Other Standards

Other useful standards include naming conventions established for company-wide use. Although this type of standardization is more of a procedural issue, it can nevertheless facilitate product development. Consider naming conventions for parts, assemblies, drawings, and features. Naming features is quite helpful during product development. Modifications and revisions can be made to specific features by simply selecting them by name.

Summary

Requirements for successful completion of a product design include breaking down the development, avoiding the cut and paste method, keeping it simple, and developing standards. These steps contribute to developing an effective blueprint, and thus are key to successful product development.

Going Beyond the Basic Sketcher

Introduction

Building flexibility into a sketch is essential. The construction and information tools in the Sketcher are also important in establishing your blueprint. In this chapter we focus on practices and tools within the Sketcher that are often overlooked, why sketches do not regenerate, and the internal workings of the Sketcher (Sketcher flow chart).

Building Flexibility into a Sketch

You can easily build flexibility by using Sketcher functions to promote desired actions and reactions in completing your product development. *Dimensioning* is the most widely used function in the Sketcher to accomplish flexibility. However, the key to effective dimensioning depends on the results you desire. In short, always dimension your sketch to accomplish a particular task. Even advanced users are found inputting dimensions without regard to the overall goal or blueprint.

The Align, Use Edge, and Offset Edge functions can also help to build flexibility into a sketch. When using these functions, existing geometry or datum planes must be in the Sketcher. As in the case of dimensioning, the most effective use of these functions occurs when each is related to a goal or blueprint.

Construction Tools

Construction tools can establish markings for the Sketcher to focus on during the regeneration of the sketch, and offer additional control to accomplish tasks for complex sketches. When you use the following command sequence, Pro/ENGINEER provides the full set of Line functions. However, instead of using a line, Pro/ENGINEER will create a construction center line.

Line → Centerline

The center line can then be used for axis of rotation, symmetry, alignment, and dimensioning.

The next command sequence provides the full set of Circle functions, but instead of using a circle, Pro/ENGINEER will create a construction circle. The circle can then be used for alignment, creating diametrical sketched features, and diametrical or radial dimensioning.

Circle → Construction

Points are also considered construction tools, and are best used to focus the Sketcher. Additional focus may be necessary during sketch regeneration. Points help to solve tangency positions and end point conflicts.

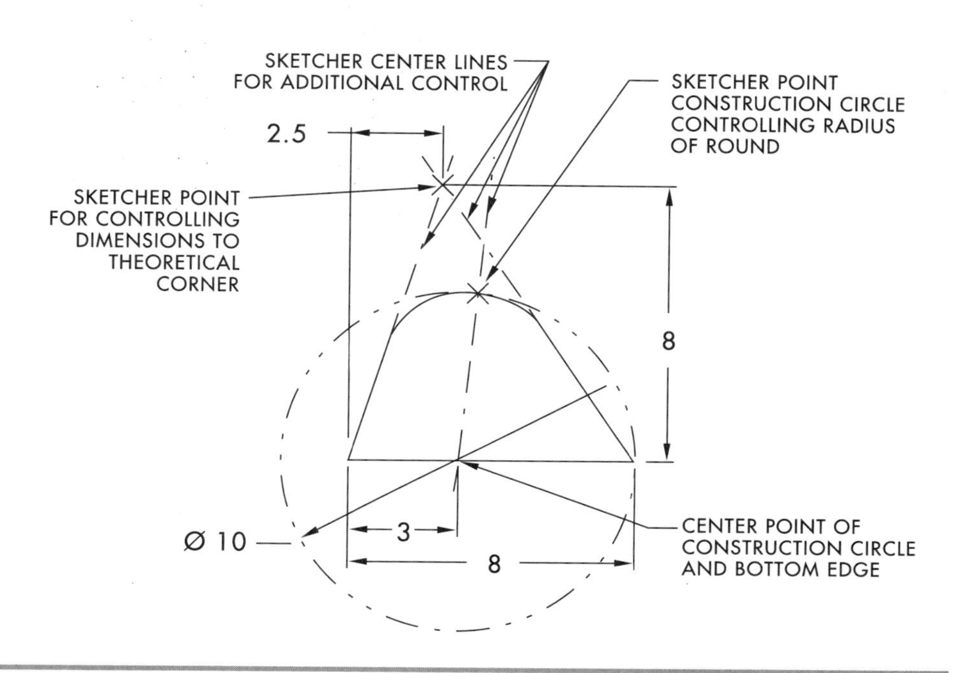

Construction tools.

Information Tools

Information tools can help you develop a sketch by providing data about Sketcher geometry. To access these tools, select **Sec Tools → Sec Info**.

Information tools serve two purposes:

- Information tools, such as the Grid Info selection, can assist you in setting up the sketch. Grid Info will give you relative spacing information on the grid which can be used as a visual reference while sketching.
- Information tools can be used to identify information specific to Sketcher entities, such as type, Angle, Distance, and References.

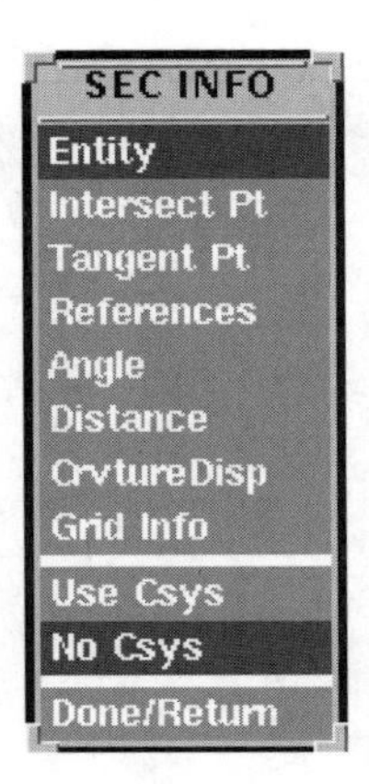

Section Information menu.

✘ **TIP:** *The Reference option is an excellent tool for redefining an already created section. Reference will color code any type of geometry referenced by the sketched entities.*

Other Tools

Additional tools or functions that can help you capture your blueprint while sketching are described below.

- ❑ Known Dimensions can be considered a tool. This function can be used in any sketch to capture information about background part geometry. Known Dimensions can also be used in Sketcher Relations to capture the Sketcher design intent as it relates to existing geometry.
- ❑ The **Geom Tools → Move** command provides you with the ability to execute dynamic "what if" studies while in the Sketcher. The Move command allows you to dynamically move any Sketcher entity or group of Sketcher entities.

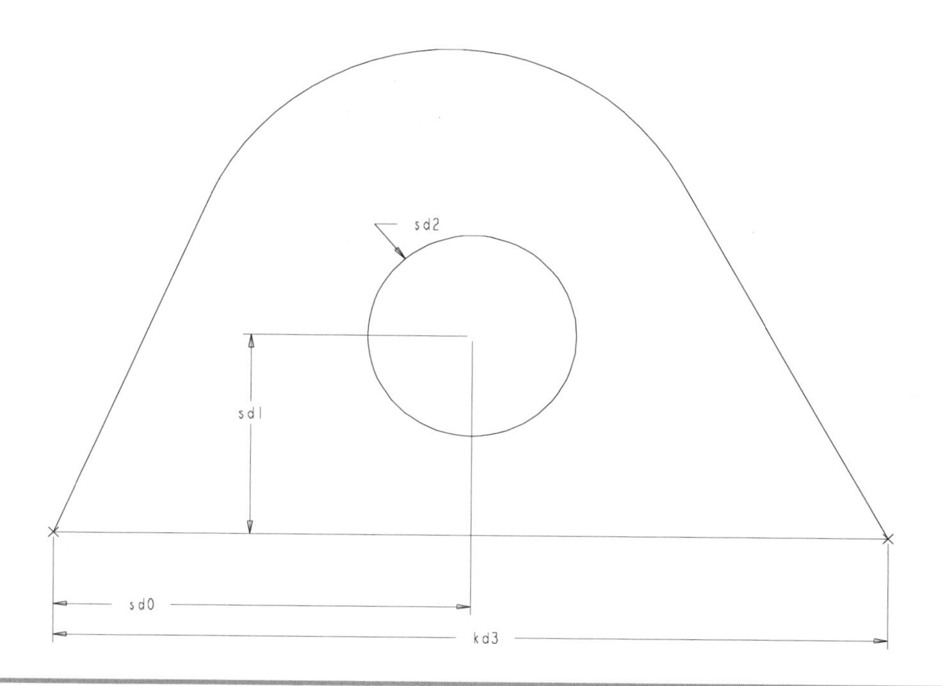

Known Dimensions in Sketcher.

✔ **NOTE:** *When you execute the Move command while in the Sketcher, a dragged entity aligned to part geometry will become unaligned. However, all dimensions already created will automatically update to the new dragged position.*

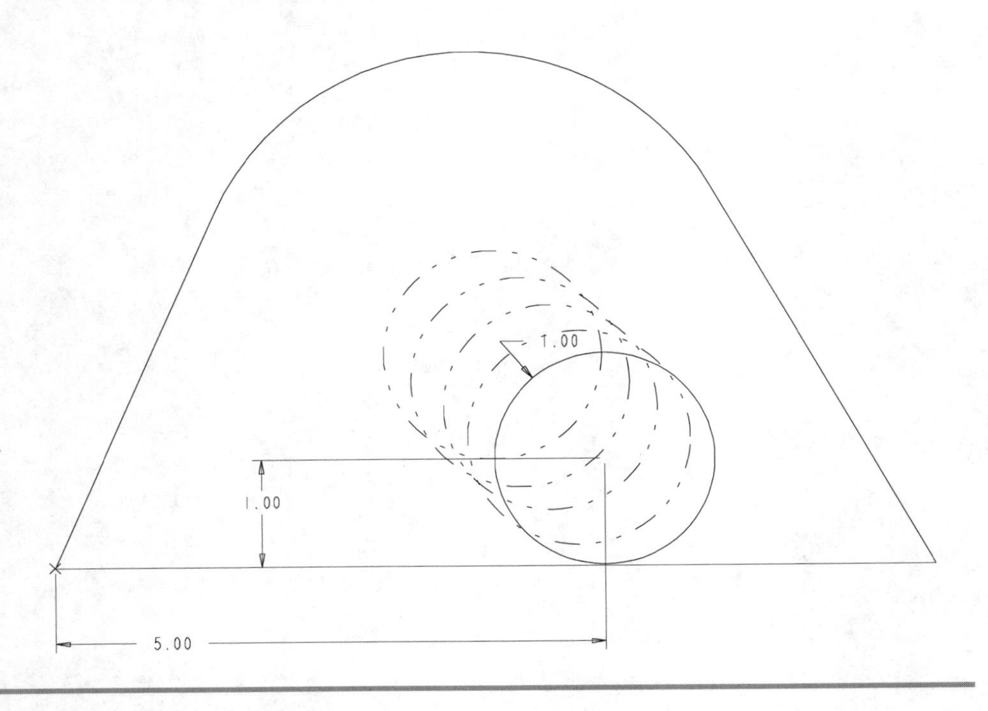

The Geometry Tool → Move command.

The **Geom Tools → Thicken** command is available when creating a sheet metal part. The Thicken command will automatically create a phantom offset line representing the thickness of the sheet metal part. This tool provides a graphical representation of how the material will develop when the feature is actually created.

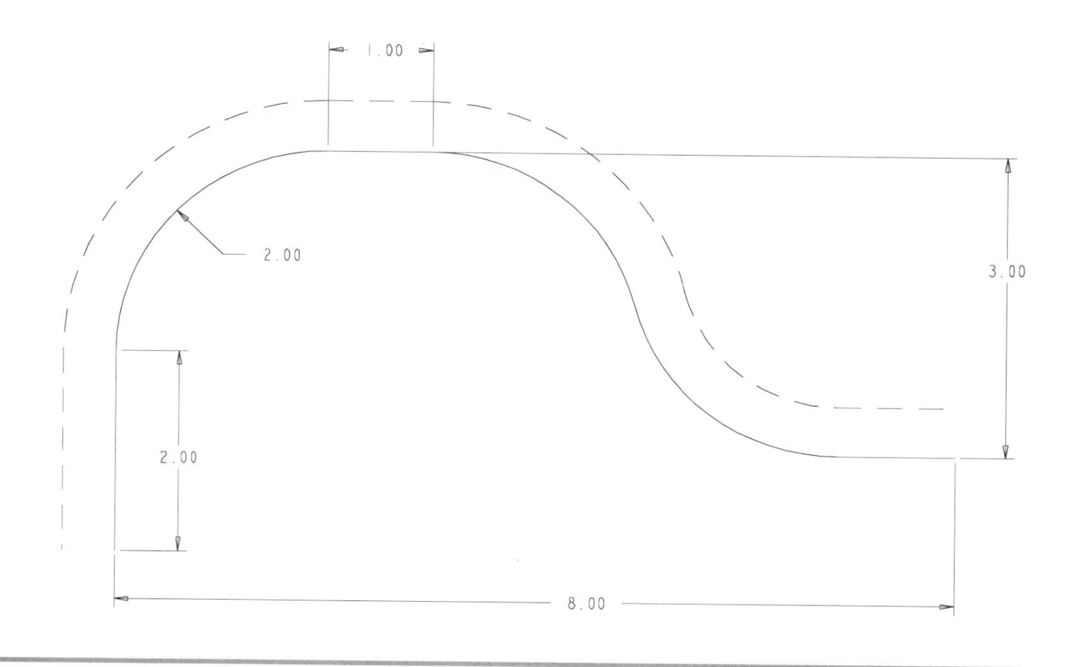

Geometry Tool → Thicken (sheet metal) command.

When Sketches Do Not Regenerate

Sketch regeneration failure may occasionally seem mysterious. The cause of a failure is in principle quite simple: at least one entity has not been defined. The following procedures are useful in identifying the cause of a failure.

- ❑ Similar to breaking down the design when developing a blueprint, a sketch can be broken down when regeneration does not work. Work your way around the sketch by focusing on one vertex at a time to ensure that each entity is defined.

✘ **TIP:** *When working with a complex sketch, you can sketch and regenerate pieces of the total. Add Sketcher entities and regenerate these portions separately until all entities have been regenerated.*

- Sketcher accuracy, or how closely each Sketcher entity is scrutinized, plays a significant role in sketch regeneration. The simplest way to overcome Sketcher accuracy is to use Zoom Out or Zoom In rather than modifying accuracy.

The Sketcher Flow

Understanding where the Sketcher fits in Pro/ENGINEER as well as the Sketcher's internal structure will help you to go beyond the basic Sketcher. The following flow chart demonstrates how the Sketcher works. Note in the flow chart how Redefine and Regenerate come into play.

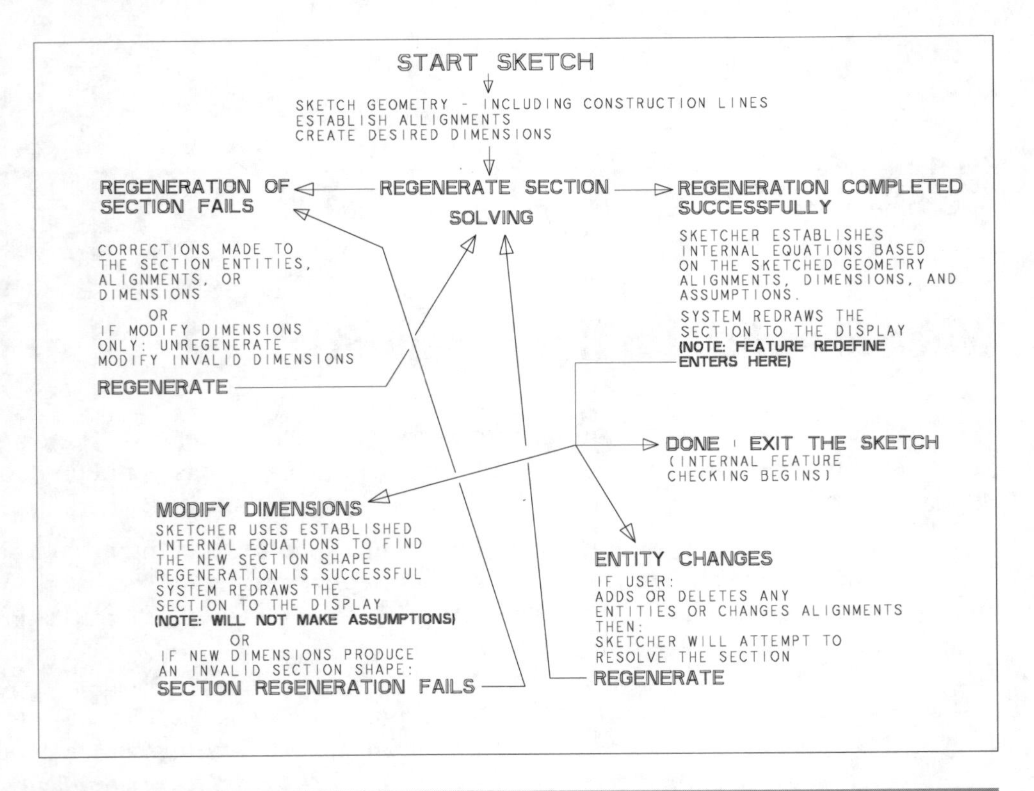

Sketcher flow chart.

Summary

The Sketcher is a powerful tool when used to the fullest extent. Remember to focus on building flexibility into your sketches. Dimensioning, Alignment, and Use Edge are examples of tools or practices which should follow your design blueprint. Other tools such as Move and the set of Section Information tools are also useful in the sketching process. Take advantage of the full suite of tools available and move beyond the basic Sketcher.

PART Two

Product Development: Part Mode

Following a Blueprint into Part Mode

Introduction

The most important aspect in developing parts is following the design blueprint. Breaking down the design should be your primary focus when creating parts. The objective is to develop parts that contain logical setups for downstream or concurrent applications including drawings, assemblies, and manufacturing. In addition, parts should be flexible, which means that they can be easily changed or modified.

Setup and Concurrent Engineering

Setting up a part while considering all aspects of product development can help users achieve *concurrent engineering*. Concurrent engineering is another term for product development when all disciplines work together virtually simultaneously.

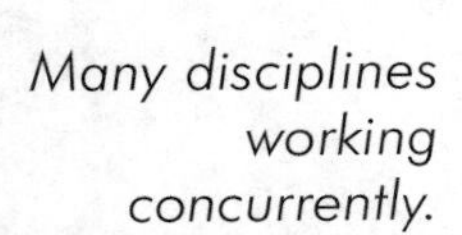
Many disciplines working concurrently.

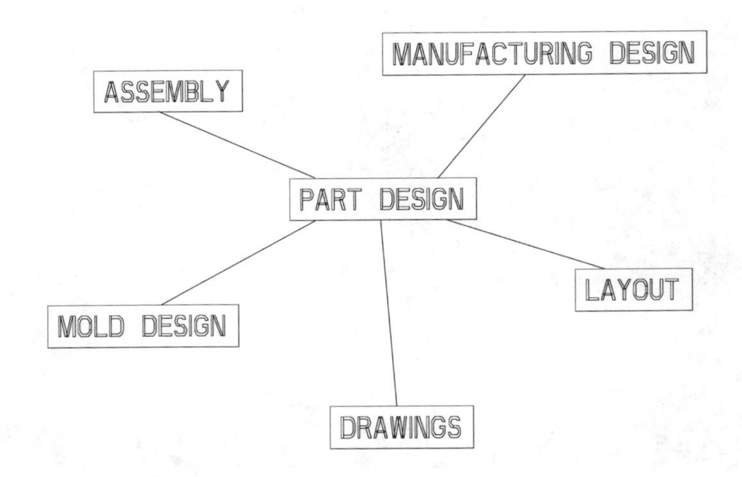

Following your blueprint into part mode is especially important because of Pro/ENGINEER's single database structure. The single database structure links all disciplines directly to the part. Therefore, as you make changes to the part in a particular discipline, all other disciplines react accordingly. As you follow your blueprint into part mode, the following simple procedures will help you accomplish concurrent engineering.

1. Whenever possible, part feature development and parent/child relationships should be reviewed by all other disciplines involved.
2. When dimensioning part features or reference geometry (datum planes and datum points), create the dimensions so that they can be shown on drawings rather than created. This simple tip can save a lot of time when creating the detail drawing.
3. Create a feature with the understanding of whether or not it will be used for assembly. Think ahead about how you wish your part features to act and react in a system or assembly. If the feature is to be used as a key feature for assembly, additional criteria may be important.
4. If you wish to mate or align a surface of a part in an assembly, the surface must be planar.
5. If you wish to insert a part feature in an assembly, the surface must be cylindrical.

6. Features should be developed for ease of manufacturing. Understanding of whether a particular feature is important or necessary for manufacturing is crucial during part development. Examples of thinking ahead for manufacturing appear below.

- ❑ In the case of a draft feature for a mold, you need to create the draft angle suitable to the material.

Manufacturing draft requirement.

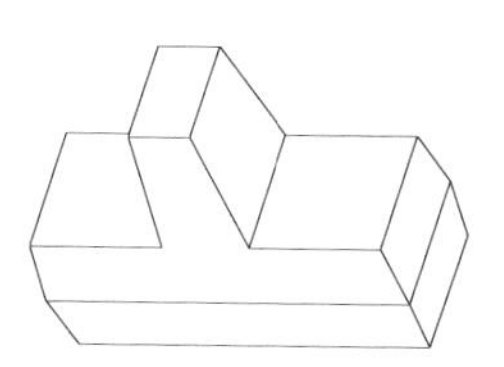

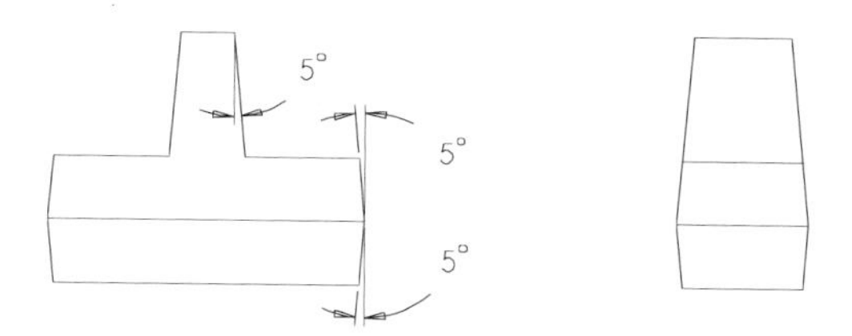

- ❑ Create a punch in the proper direction for stamping.
- ❑ Verify that the smallest round feature created is equal to the smallest numerical control cutter tool.

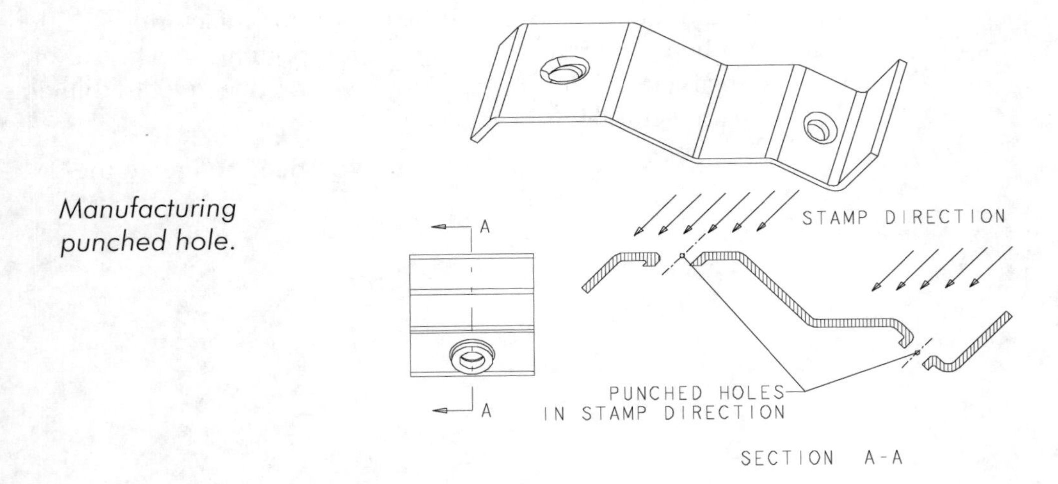

Manufacturing punched hole.

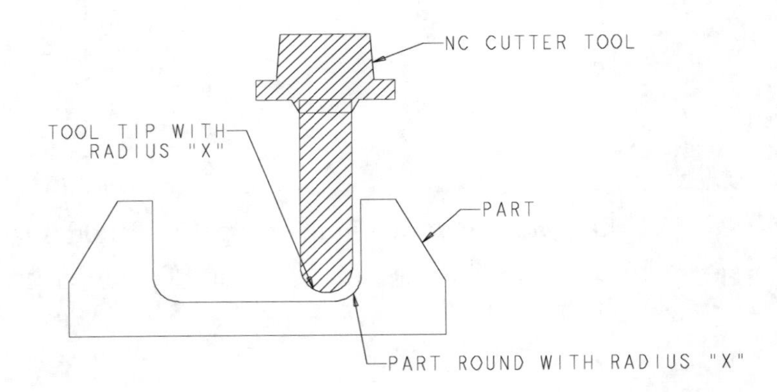

Manufacturing cutting tool related to round size.

The above examples point out the importance of thinking about other disciplines in the interest of streamlining product development. In addition, you will get what you need and expect during development, thereby avoiding costly surprises and changes. This practice is beneficial only if the design blueprint or development intent is adhered to.

Flexing the Part

One of the best tips for any user is to flex the part during development to ensure quality. Flexing the part is simply the act of making modifications and changes to the part. A quality part is one which fulfills requirements of all disciplines. In addition, a quality part can be easily changed through redefine or modify. In the course of flexing the part, problems or failures may occur. Part failures are beneficial in that they point to problems that may not have been noticed during blueprint development. Failures give you an opportunity to fix the design up front during development rather than later when fixes may be difficult. Basic guidelines for flexing a part include the following:

- Redefine or modify to verify whether the part will act or react according to the design blueprint (intent).
- Play the "what if" game. As you add part features to your model, consider different design scenarios that may occur.
- Redefine or modify values according to different design scenarios and Regenerate.
- Make flexing modifications after every 10 to 12 features and regenerate.

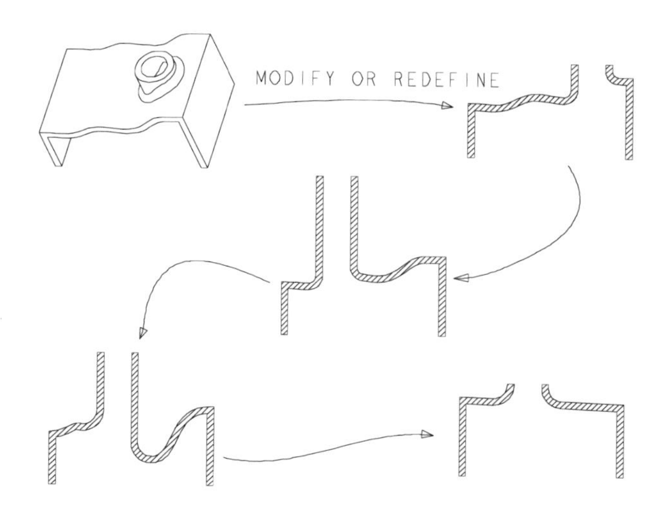

Flexing example.

Summary

During part development, the most important item is the product's design blueprint or intent. When a focus is set up, established and maintained, remarkable things will take place. Other applications can be run concurrently and costly surprises or changes during product development will be limited. Flexing a part also aids in part development. Flexing will help ensure design intent and point out possible problems in creating the part. Coupling focus and flexing will result in a quality part that is easy to modify and change, and that fulfills your design blueprint.

Complete Part Creation Using the Shell Function

Introduction

The shell feature is one of the most widely used functions within Pro/ENGINEER. Although many users have used the shell feature when creating products such as plastic injected molded parts, even experienced users have not exploited the shell feature's one-step function to create a complete part. You can rapidly develop complete parts by using multiple surface selections and parameters to define wall thickness in concert with stamping or mold design rules. The shell function is especially useful for creating parts such as brackets and flanges. A simple example, shown in the following illustration, is a block made into a thin-walled bracket.

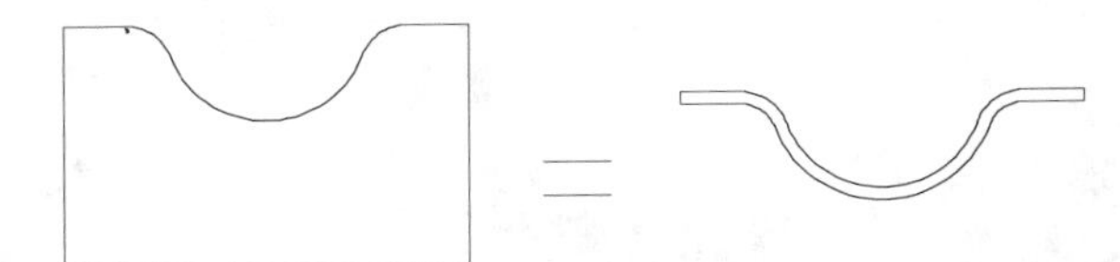

Simple representation using the shell function.

This chapter will focus on setting up a part to be created using the shell feature. We use a sample problem illustrating how to plan, set up parameters to control wall thickness, and verify the shell during development. The part is also subjected to prototype testing for structural and packaging verification, and changes will be incorporated based on this testing. Changes to the shell will include adding strengthening ribs and a clearance hole to meet additional design requirements that were not originally considered for the shell. Results will demonstrate how to take advantage of the shell to reveal a complete part.

Completed part using the shell function.

Planning Shelled Parts

Creating shelled parts can be quite simple once the proper blueprint is established. Preliminary planning for the basic elements of your design will reveal numerous options during product development.

The first step in developing a part using the shell function is to determine whether the part will be designed to the inside or outside of the material. This development discussion should be based on engineering and manufacturing department requirements. The process is very similar to the development of a mold base or even a die.

The most important reasons for beginning with this step are to eliminate unnecessary features and to make the design more flexible. If your design demands that most of the dimensions for product development are defined to the inside of the material, Pro/ENGINEER's shell feature would be used with a negative value. In contrast, if your design requires that most dimensions be developed to the outside of the material, you would use a positive shell feature. This approach will allow drawings, gauge fixture requirements, and manufacturing requirements to "fall out" of the design based on the inside/outside requirement.

A negative value will force the surfaces of your design to offset outward rather than inward with the typical positive shell value. The following illustration shows a negative and a positive shell.

✔ **WARNING:** *The decision to create a negative or positive shell is important because you cannot redefine the shell to the opposite direction in which it was originally created.*

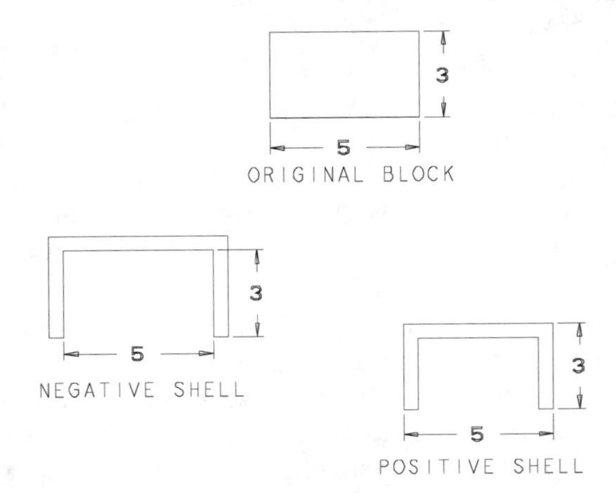

Negative and positive shell values.

Creating a Stamped Metal Part

A common use for the shell feature is the creation of a thin-walled stamped metal part. In our example, we will create a metal cover. As with any design, you begin with default datum planes and then commence building the features. Start developing the base feature by building from large to small scale.

In this example, we begin with basic requirements from manufacturing: the controlled dimensions and all radii must be developed based on the inside of the material. With these requirements in place, the shell feature takes care of the rest.

1. Before creating the first feature, set up a parameter to control wall thickness. Assume that the parameter derives from a manufacturing requirement which dictates that wall thickness will be three millimeters thick at all times. From the Part menu select the following:

Set Up → Parameters → Part → Create → Number

2. When prompted for the parameter, key in *wall_thk*. Next, when prompted for the value of *wall_thk*, key in *3*. Finally, select **Done → Return.**
3. The first feature of the metal cover is a revolved thin section on the bottom side of datum A.
4. Add another feature to create the dome portion of the part. A blend of two sections creates this shape.
5. Create rounds to satisfy the development requirements for inside radii.
6. Create the shell by selecting the bottom flat surface of the metal cover for removal. Use a negative shell to force the finished material to the top side of datum A. This will create a flat surface for quality control to use as the primary datum to gauge the part.

Before and after the negative shell.

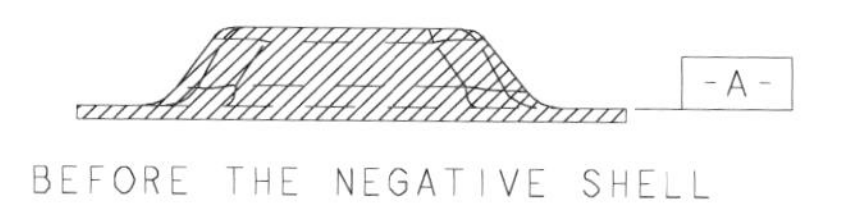

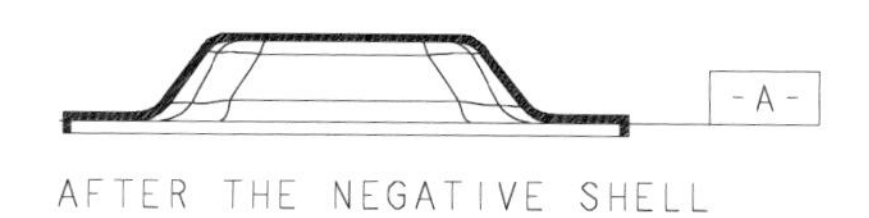

---

7. Verify the part design by creating cross sections and visually inspecting them. Check for potential geometry problems.
8. Look for geometry that does not follow your design blueprint (design intent). The shell feature created a lip around the perimeter of the part, which is not indicated in our design intent.
9. The shell feature will require two changes: a modification to the first feature created, and setting the thickness of the revolved thin wall equal to the parameter called *wall_thk* (wall thickness). In addition, the shell feature should be redefined by adding additional surface removals (break out surfaces).

Undesired lip created during shelling.

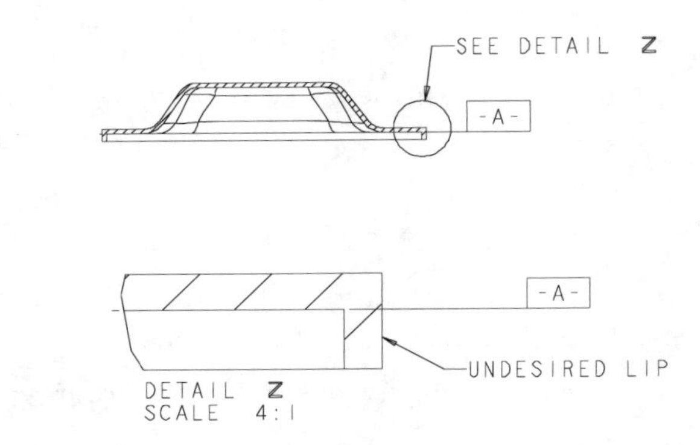

Modifying the first feature's value equal to the part parameter wall_thk.

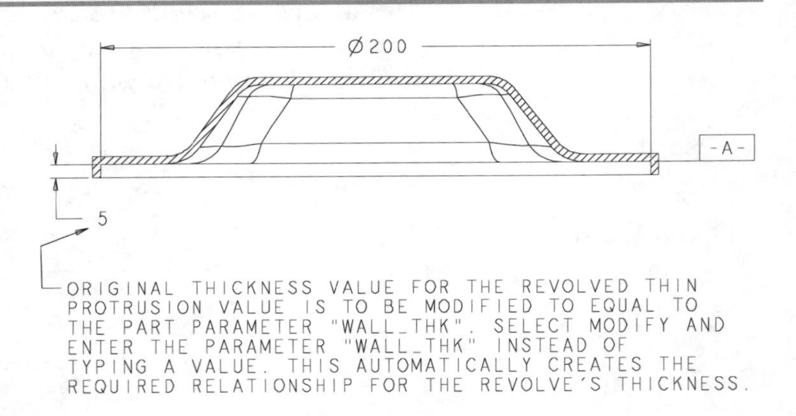

Additional surfaces to be removed (break out surfaces).

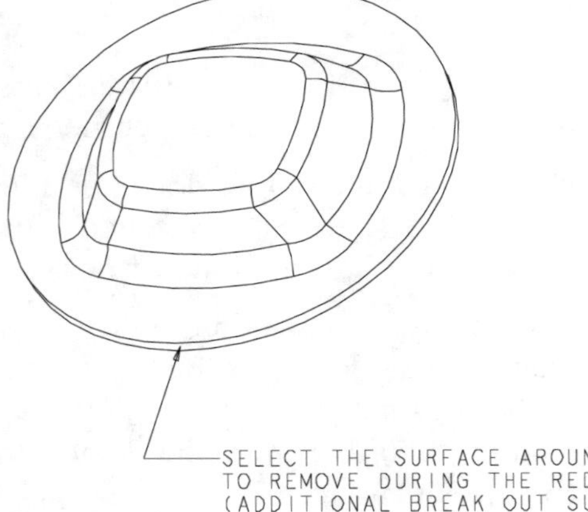

The part is now complete for prototype testing.

Shelled metal cover for testing.

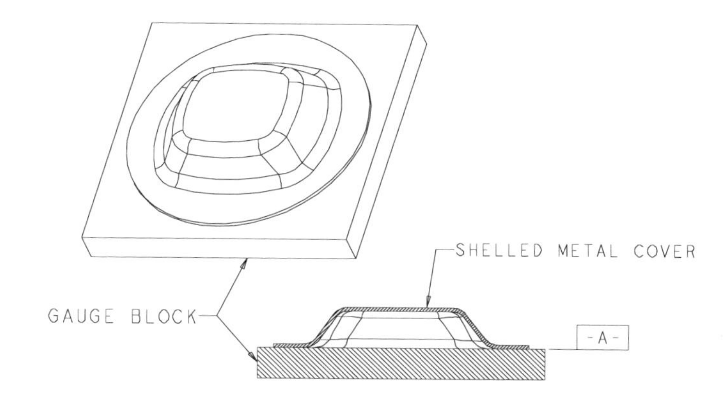

Changing the Shelled Part After Testing

After testing is complete, additional changes to the shelled metal cover are required. Assume that the prototype analysts recommended that four strengthening ribs should be stamped into the part to stiffen and strengthen the structure, and that a clearance hole is needed at the top of the part to facilitate assembly.

To create the strengthening ribs of the metal cover, we will use the Insert function to place the features before the shell. By using Insert mode, the strengthening ribs can be developed based on our original design requirements. By canceling Insert mode when the ribs are complete, the shell feature will return, automatically developing the new metal cover with the additional information.

After placing the rounds, add the rib features. Activate Insert mode from the Part menu by selecting the following commands:

Feature → Insert Mode → Activate

Insert location for strengthening ribs.

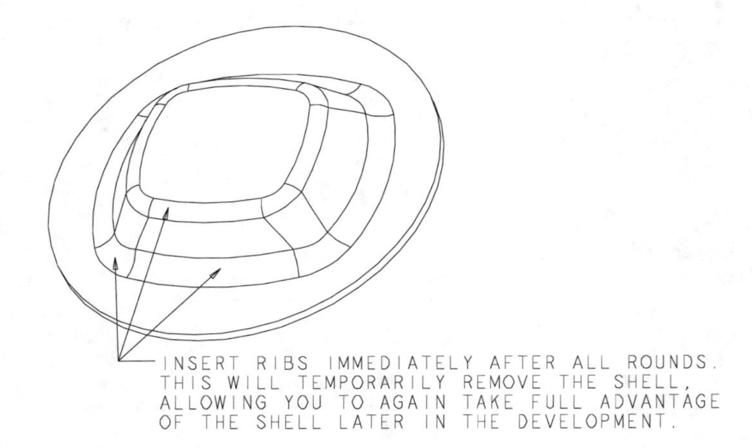

INSERT RIBS IMMEDIATELY AFTER ALL ROUNDS.
THIS WILL TEMPORARILY REMOVE THE SHELL,
ALLOWING YOU TO AGAIN TAKE FULL ADVANTAGE
OF THE SHELL LATER IN THE DEVELOPMENT.

There are many options available for creating the strengthening ribs on the shelled metal cover. Our goal is to create parts that are easily changed and modified. The most important requirement is that added features must take advantage of the shell feature's ability to generate a complete part.

The additional ribs will be patterned and follow the contour of the part through the use of a variable section sweep. All sharp edges will be removed by using rounds to allow for the shell to properly offset all surfaces and to create the completed shelled metal cover.

We will use a projected curve with an angle to start the development of the strengthening ribs. The curve is developed by sketching on datum A and projecting the curve to the top surfaces of the metal cover.

✔ **NOTE:** *The horizontal reference plane for the sketch of the projected curves should be created at an angle on the fly to allow for easy patterning.*

1. Sketch the curves to be projected. Although the projected curve is a single feature, the sketch will contain two sketched lines. This will avoid the creation of an extra projected curve to control the variable section sweep, thereby reducing the total number of features used in the part.

2. Select all surfaces on the top of the part for the datum curve to be projected onto.

✔ **NOTE:** *As the design changes, surface selection can be easily redefined to add or remove additional surface references for the projected curve.*

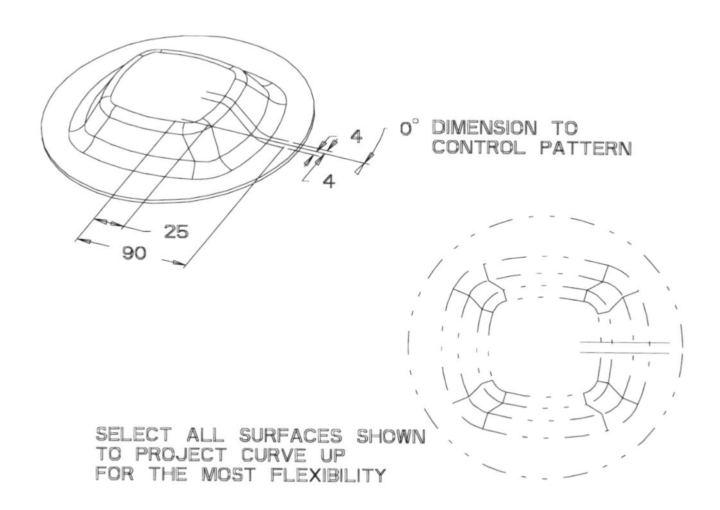

Projecting curves and selecting surfaces.

3. Create a variable section sweep by selecting one of the curves as the spine and the other as the horizontal vector for the sweep. After finishing these selections you will be placed in the Sketcher with the cross hairs showing you the way for the sweep.
4. The sketch consists of a simple three-point arc with the ends lying on the cross hairs. The width of the curves will control the actual radius of the rib and allow for additional changes should they occur. More importantly, the end points of the arc follow the exact contour of the metal cover.

Sketch view and resulting sweep.

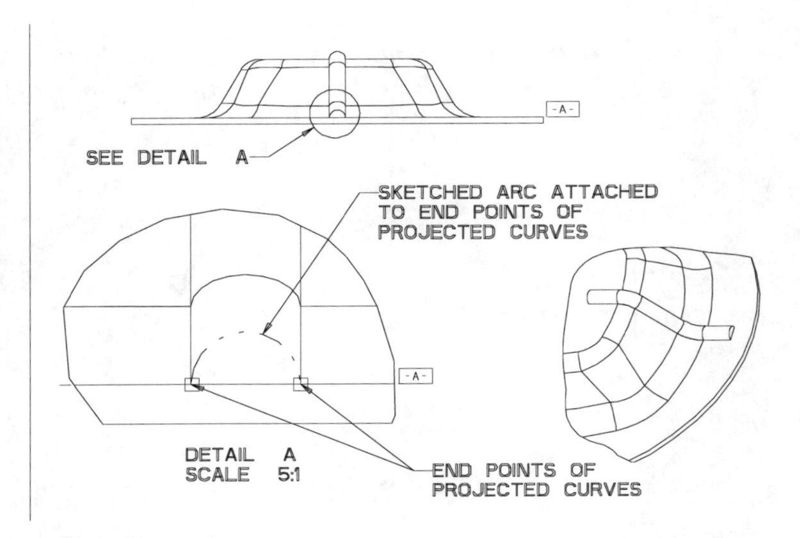

The next steps require focus on additional design requirements. Because the part is to be stamped, the absence of sharp edges is important to prevent tearing or weakening of the material.

1. Create rounds on each end of the sweep.

End rounds.

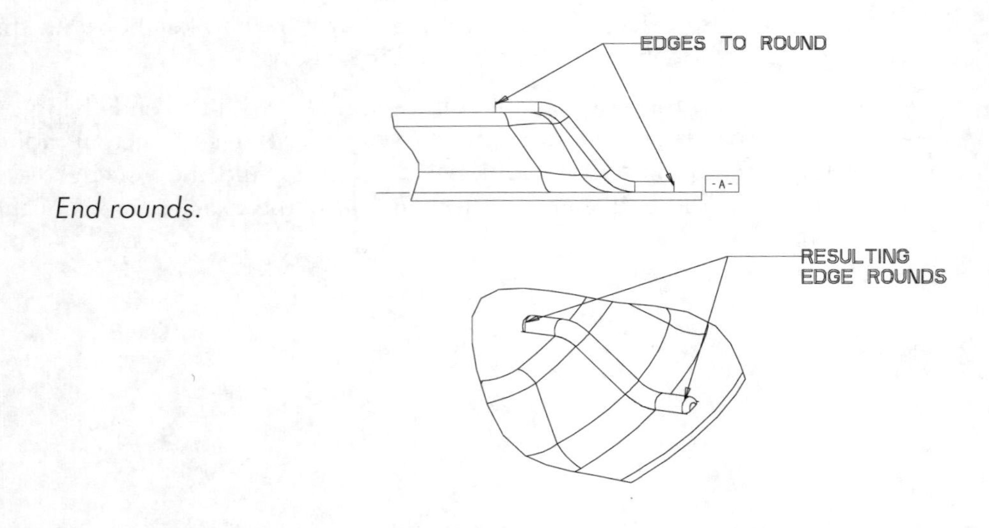

2. Round the complete outer edge of the strengthening rib by using the Surface to Surface option. The surface-to-surface round will allow for greater flexibility since the sweep extends over multiple surfaces. In addition, a surface-to-surface round allows for the removal of surface patches.

✔ **NOTE:** *Rounds and all features for the sweep will be included in the pattern.*

Recall that the design intent also dictates that the sweep and all rounds will be included in the pattern. When selecting surfaces for the surface-to-surface round, select surfaces that will be similar regardless of where the rib is swept or patterned on the metal cover. Otherwise, the rounds may fail. Rounds are like any other feature within Pro/ENGINEER: thinking about how you want them to work will significantly benefit you in the long run.

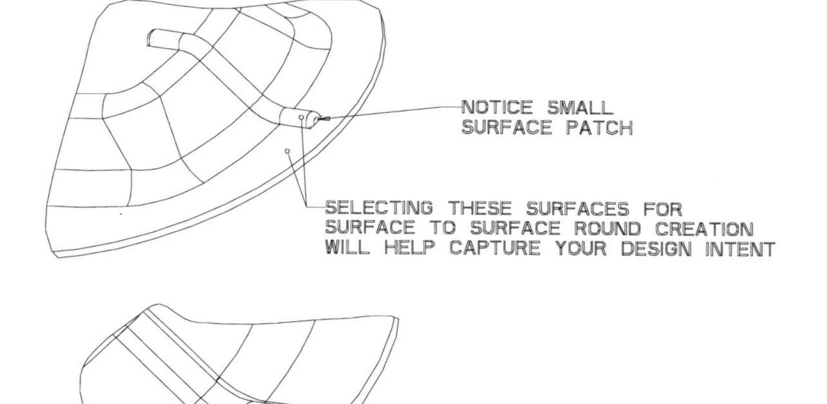

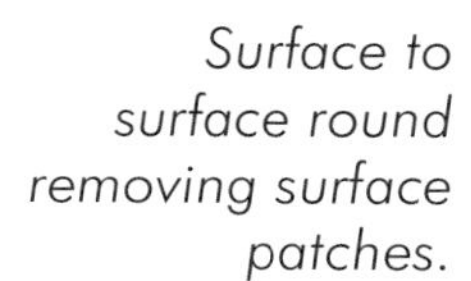
Surface to surface round removing surface patches.

The final step is to pattern the rib which consists of four features. To create the pattern, take the following steps:

1. Create a local group (Local Group option in the Group menu), and give it a logical name (e.g., *strength_ribs*).

2. Pro/ENGINEER will prompt you to select the features to be grouped. Select the first feature of the group (projected curve) and the last feature (surface-to-surface round).
3. When you are prompted on whether you wish to group all features in between, answer Yes. The local group is created unless the features are not in sequence.
4. In the Group menu, select the Pattern option for groups. Select a point near the group. Pro/ENGINEER will automatically grab the group for you because the program will seek only a group.
5. Pattern the group about the angle dimension from the projected curve which was created on the fly. After you input the number of ribs, the ribs are complete.

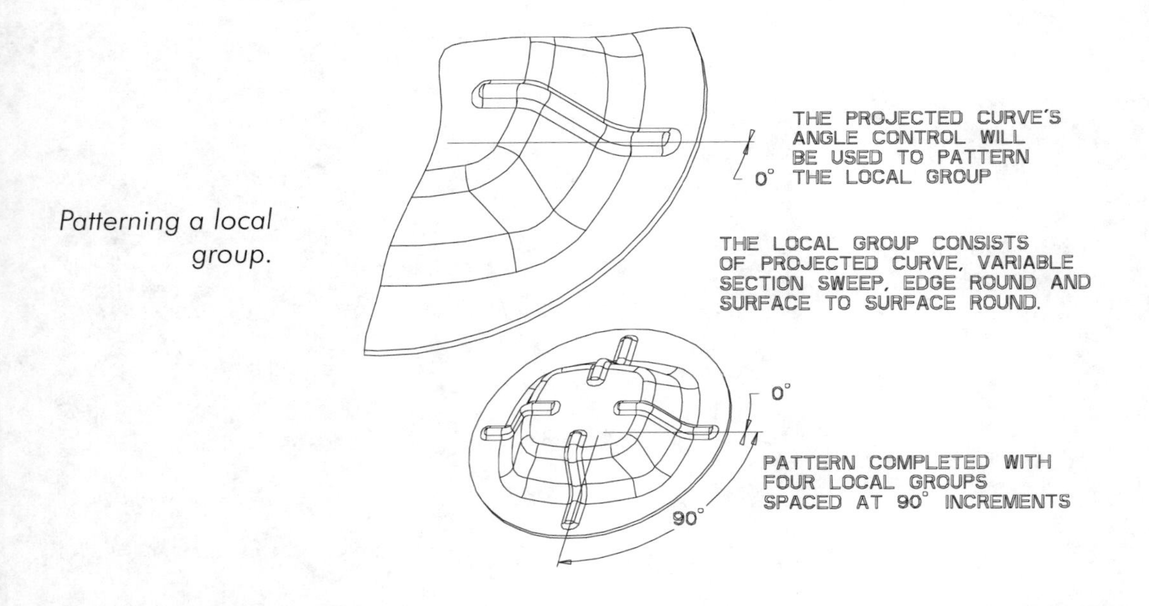

Patterning a local group.

Keep in mind that the ribs have just been created using the Insert function, and that the shell feature which automatically completes the metal cover is still there. Before you create the clearance hole, cancel the Insert function and examine the shelled metal cover.

Shelled metal cover with strengthening ribs.

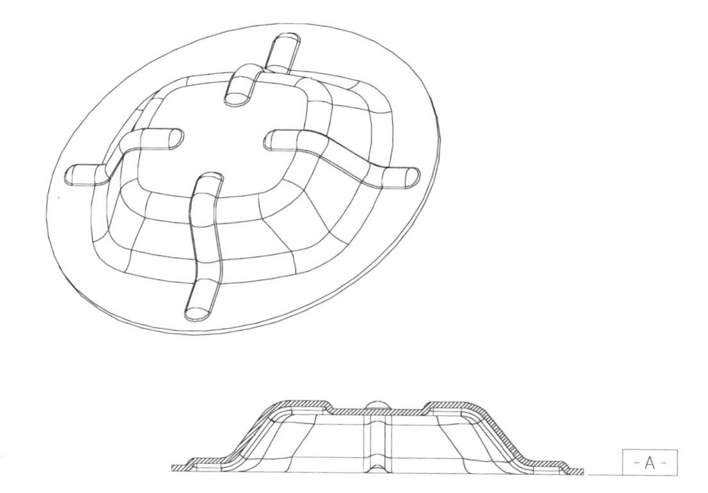

Changing the Shelled Metal Cover: Adding a Clearance Hole

The clearance hole may also have certain design requirements which expand the original design blueprint. In this case, the hole in the shelled metal cover must be at the center of the part and dimensioned to the outside of material to allow for assembly access. This problem poses two issues: (1) how to incorporate the feature or features and still take advantage of the shell feature, and (2) the need for a dimension to be created to the outside of material.

Using the Insert function in conjunction with Part parameters and Sketcher tools will allow you to follow the design blueprint and capture exactly what is required. The first step is similar to creating the ribs. Determine where the clearance hole should go in the sequence of developing the metal cover. We already know that the clearance hole should precede the shell feature so that after the hole is created, the shell feature's one-step function can create the finished part. The clearance hole will probably not change much because it is necessary for assembly. Because changing the hole is improbable, and because of its placement in the center of the part, this feature will be inserted early in the part development sequence. Consequently, the feature will be inserted immediately after the second protrusion (blend).

1. Create a hole blind depth coaxial to the first revolved feature.
2. Place the feature on the top surface of the part.
3. When prompted for the diameter, enter the desired value plus two times the wall thickness. Use the *wall_thk* part parameter when inputting the following value: *(10 + (2*wall_thk))*. This will create a relationship and ensure that a proper clearance hole will be made after the shell is resumed from the Insert function.

Inserted blind hole with value tied to wall_thk part parameter.

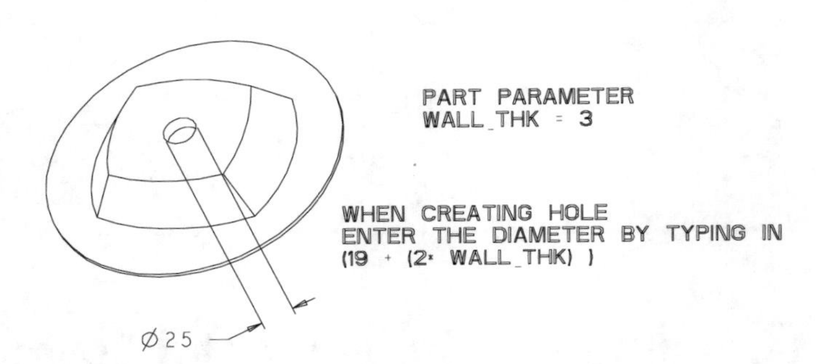

4. Another option for the clearance hole would be to create a blind cut similar to the hole. The blind cut is inserted right after the second protrusion (blend).
5. While in the Sketcher, use construction tools (Construction Circle) to develop the proper sizes. This method is used so that a dimension can be easily shown on a drawing without creating a dimension.

Cut using Sketcher construction tools and wall_thk part parameter.

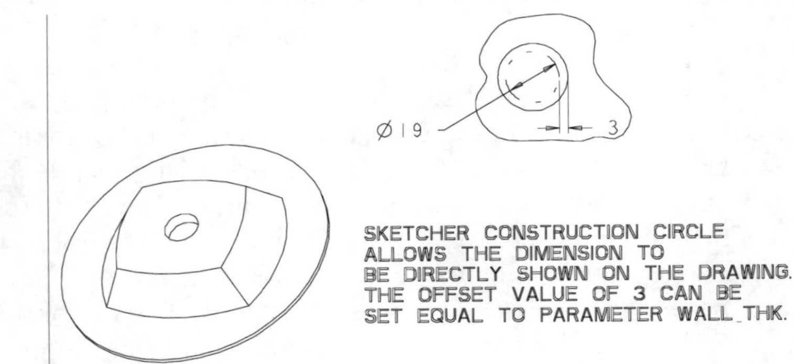

After finishing the hole using either option, add round along the top edge to smooth out the part. When complete, cancel the Insert function. The shelled metal cover is complete, with one exception. The final design requirement is to redefine the shell and remove an additional surface (break out surface) at the bottom of the clearance hole.

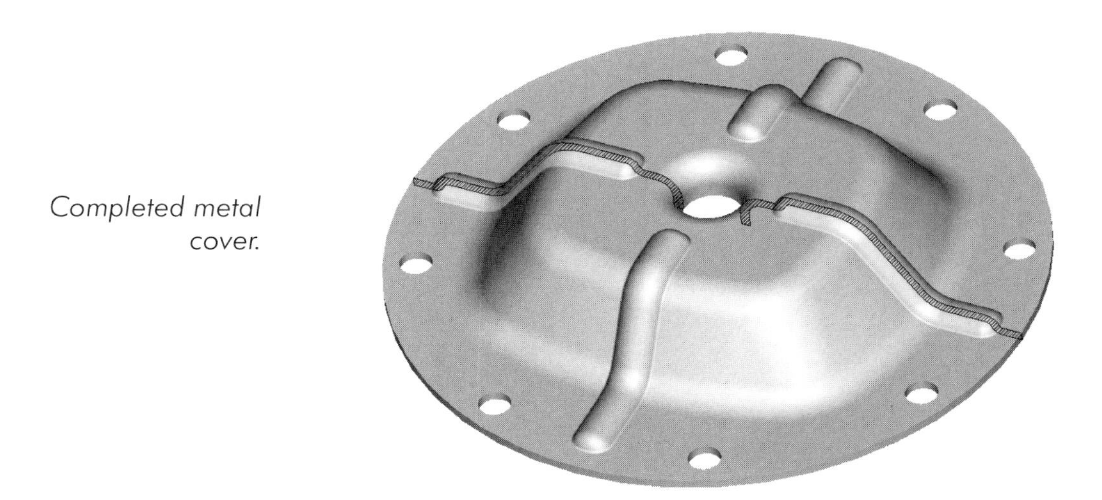

Completed metal cover.

Summary

This chapter demonstrated how developing a complete part using the shell feature can be accomplished. Setting up, planning, and verifying the shell feature as it is being built are critical stages in the development of a shelled part. Incorporating additional features to a shelled part can be accomplished by using the Insert mode before using the Shell feature. The Insert mode demonstrated how easily the Shell feature adjusts and reveals a new complete part.

Working with Non-Planar Shapes

Introduction

The typical learning process for using Pro/ENGINEER begins with protrusions followed by blends and sweeps. Some advanced users question this sequence because users often find themselves developing geometry which is not prismatic. Style and function may dictate that geometry flows smoothly to define product shape.

Your capabilities are greatly expanded once datum curves and surfaces are understood. More importantly, such understanding permits you to quickly advance to non-planar shapes. This chapter focuses on using a variety of datum curves employed in conjunction with variable section sweeps. We will explore the use of graph features and curves developed from equations to aid in developing non-planar shapes.

Development Core: The Blueprint

Datum curves and surfaces can be used as the blueprint in developing complex shapes. The base features, datum curves and surfaces, become the core of your development. In addition, the curves and surfaces are often simple to develop and quick to regenerate. An example would be creating three simple two-dimensional curves and a variable section sweep (multi-trajectory) along the three curves rather than creating a blend feature.

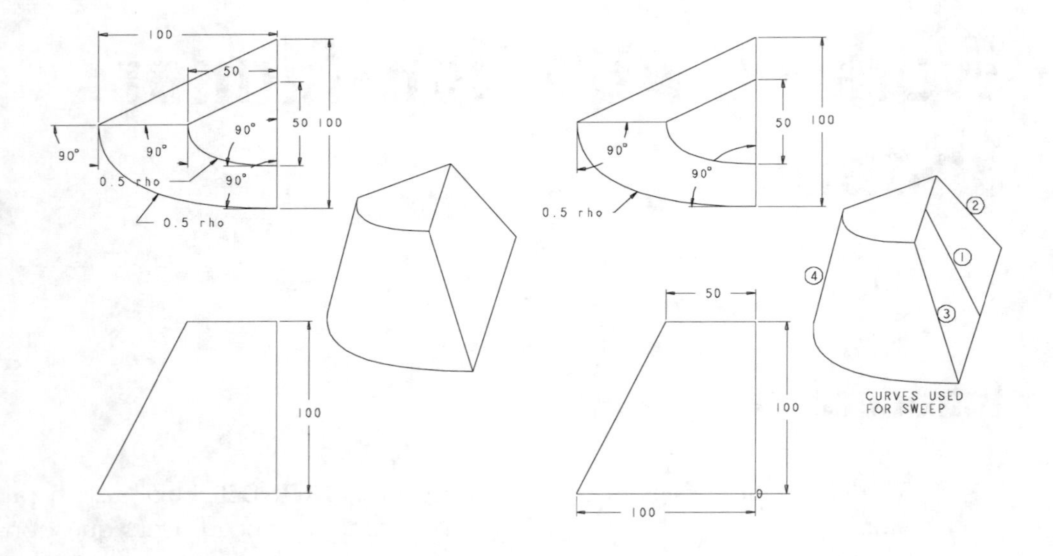

Variable section sweep versus blend.

The sweep will frequently regenerate faster than the blend, even in the simple case illustrated above. With the sweep, everything is fully defined by a series of simple features. In the case of a blend, Pro/ENGINEER needs to precisely calculate what the geometry should look like between each cross section of the blend. Although the blend feature is extremely powerful, advanced users opt for the variable section sweep whenever possible.

Recall that you should always focus on what your blueprint requires. This type of development is also very flexible. Because the development is based on simple features, any change to a simple feature will "ripple through" the design blueprint without the need for complicated cut and paste procedures. The use of simple features imposes no limitations in that they can be used in building complex designs.

Developing a Laundry Detergent Bottle

Product development of a laundry detergent bottle will be used to illustrate the power and flexibility of using datum curves and variable section sweeps to create non-planar geometry. You can create and control complex variable section sweeps (multi-trajectory) in many different ways. For instance, variable section sweeps can be developed relating to planar surfaces, datum planes, and two- and three-dimensional curves. Next, you can use graph features to control any number of parameters within a variable section sweep. This type of control makes the variable section sweep a potent tool for advanced users. Like any other design, setting up a blueprint is important because it defines what we need and how we want to get there.

Assume that the marketing department wants a laundry detergent bottle with a strong, rugged appearance. Marketing has already submitted a sketch of the basic desired shape to the design group.

Marketing department version of the laundry detergent bottle.

Laundry detergent bottle.

After reviewing product development plans with marketing, the manufacturing department expands and improves on the basic blueprint for the design. The laundry detergent bottle will have a smooth contour from the bottom to the top. All orthogonal views of the product are to have some curvature. With this basic blueprint we will begin developing the base features.

The base features consist of five simple curves defining the shape of the bottle. The first curve will be used as the spine for the variable section sweep. (The cross section stays normal to the spine as it is swept.) The first curve will also be used to set up parent/child relationships for the remaining curves, thereby controlling the bottle's height.

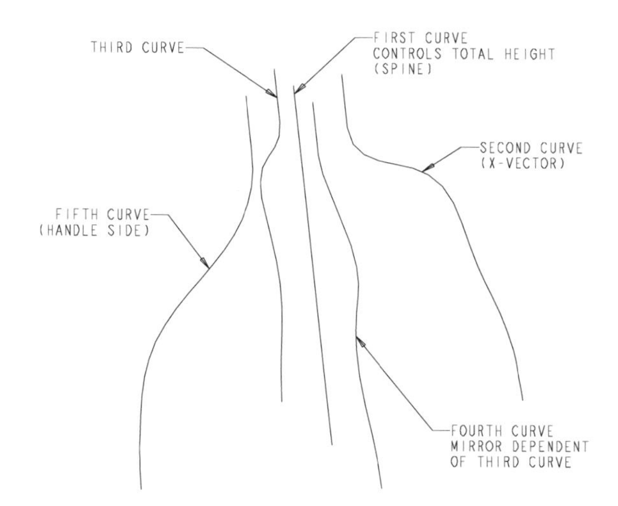

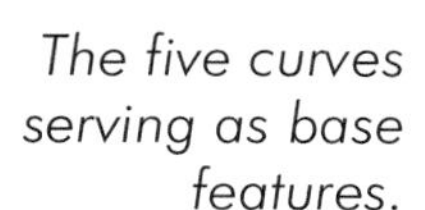
The five curves serving as base features.

Creating the shape of the laundry detergent bottle from the five curves and controlling the sweep with a spine will provide the flexibility and control needed to continue the product development.

At this point, you can create a variable section sweep along the five curves using the first curve as the spine. As with any variable section sweep, once the trajectories are selected Pro/ENGINEER will place the view in the sketching orientation for completion of the section to be swept.

To establish the basic shape of the laundry detergent bottle, create the section using two splines with tangency at both ends.

After the protrusion is complete, flex the part to verify the blueprint. Flexing the part can be accomplished by modifying the first curve and verifying that all the other curves and the variable section sweep change accordingly. Additional flexing may include redefining some of the curve shapes to verify that they generate the expected results.

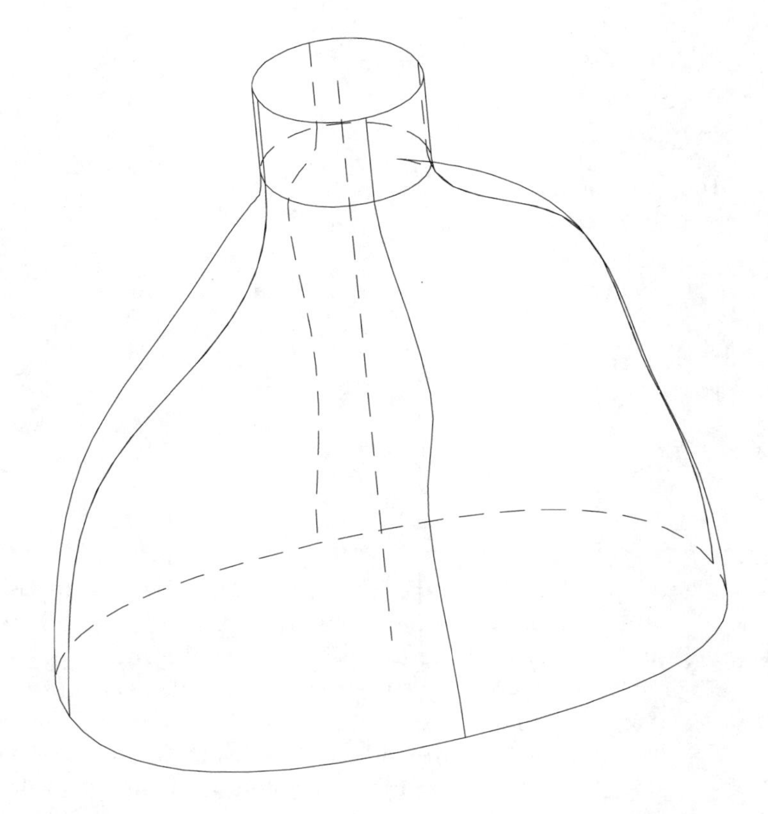

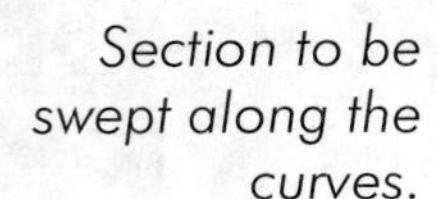

✘ TIP: *Although trajectories for variable section sweeps can be created on the fly, when creating complex shapes advanced users often prefer creating the trajectories ahead of time. In this way, the core is established. Use Insert mode in conjunction with Redefine to add or remove curves for variable section sweeps.*

Along with flexing the part, assume that you review the basic shape with the marketing department as well as manufacturing to verify the design blueprint. As with any product development, changes are typical. Because you have developed the blueprint for this part, changes can be easily incorporated.

After a quick review of the detergent bottle design, marketing approves, but the manufacturing points out a few problems. For instance, in its current shape the bottle cannot accommodate an existing screw-on lid. Use of the existing lid would be a large cost savings. In addition, manufacturing requires more curvature to the sides of the detergent bottle to facilitate molding.

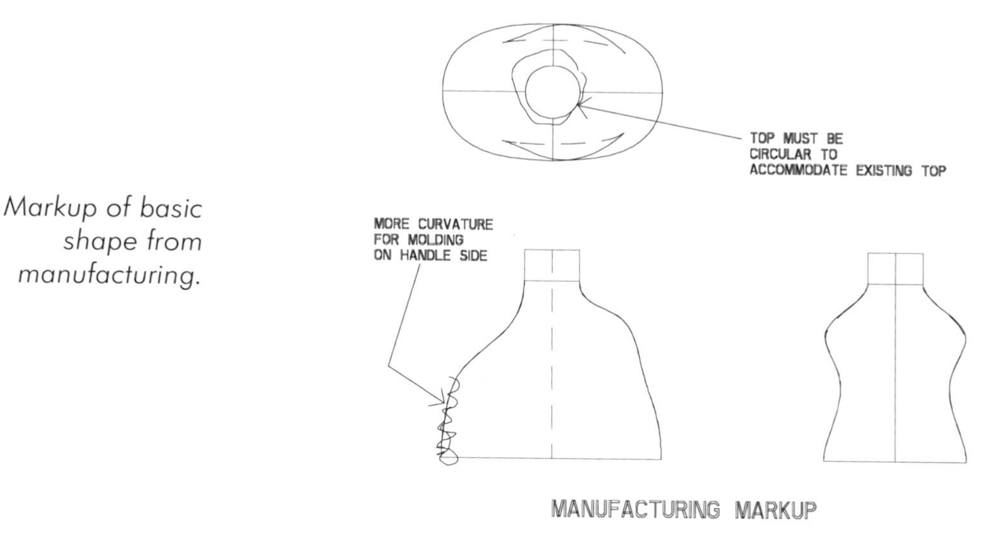

Markup of basic shape from manufacturing.

Accommodating the Design Changes

The changes required by manufacturing affect only the variable section sweep. However, at this juncture the blueprint should be considered. To carry out the design changes, take advantage of options such as Insert mode, the Graph feature, and Redefine.

The first item in the blueprint to focus on is the product development core (recall the breakdown). The curves defining the variable section sweep are the core elements of the laundry detergent bottle. The design changes must accommodate the need for a circular top so that the screw-on cap will work properly. We will also need to add two additional curves to the variable section sweep. The additional curves will resolve the molding issues.

First, focus on the need for the circular top. Changes will be made to the datum curves which control the variable section sweep.

Next, the datum curves will be redefined so that they stop short of the top. When redefining the datum curves, give special attention to maintaining the parent/child relationship for the total height, and controlling the height by the first curve. This can be accomplished by using Sketcher center lines, alignments, and dimensions while redefining.

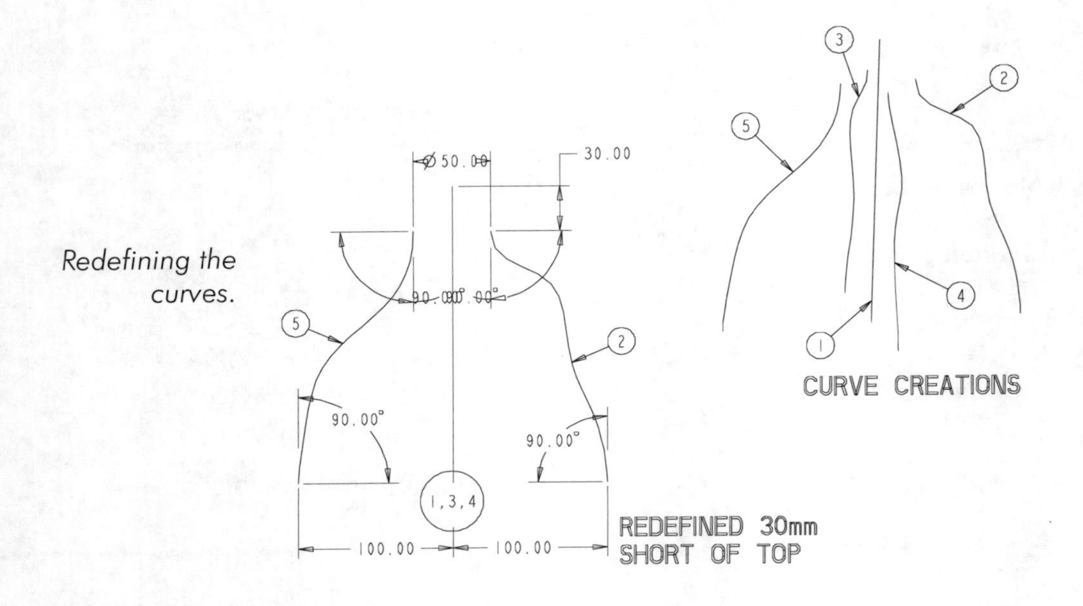

Redefining the curves.

Recall that developing parent/child relationships which yield exactly what you need is crucial to creating a flexible part. Once the redefine is complete, the variable section sweep will stop at the end of the shortest curve. This change will allow for adding material in any shape for the screw-on cap.

The next phase in accommodating the design changes focuses on adding additional curves to the variable section sweep. This change will resolve the molding issues.

The curves will be created by projecting one of the existing curves onto a surface resulting in a three-dimensional curve. To create the curve, you must first determine where the surface and projected curve should be placed in the part development sequence. Review the complete part sequence through regeneration information. The surface and projected curve will be added just after the fifth curve of the original sweep.

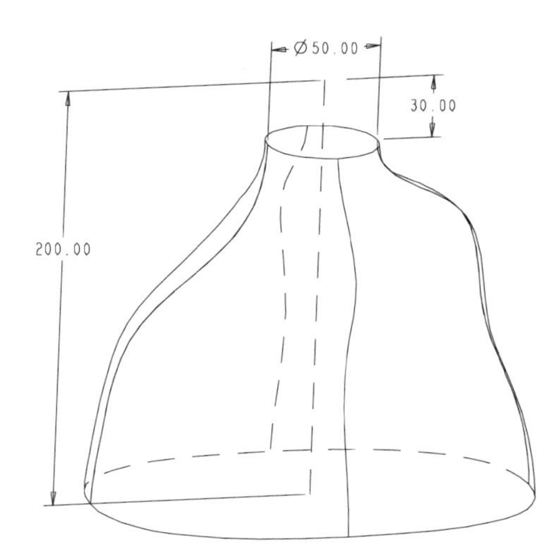

Laundry detergent bottle after redefining.

The surface to project the curve onto will be created using the simple Extrude option. The surface will have the necessary curvature required by marketing, as well as tangency at both ends.

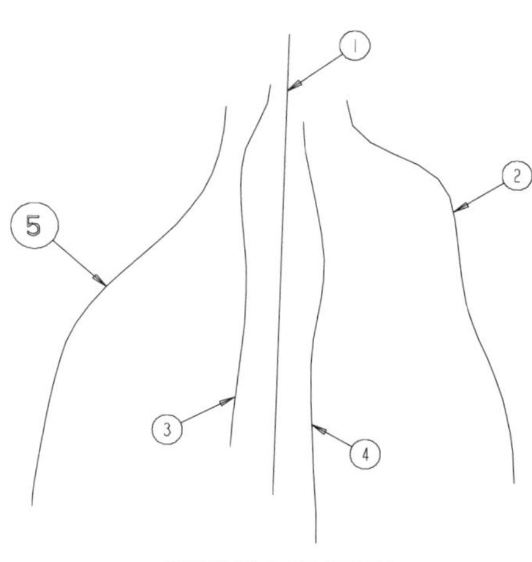

Insert location for surface and projected curve.

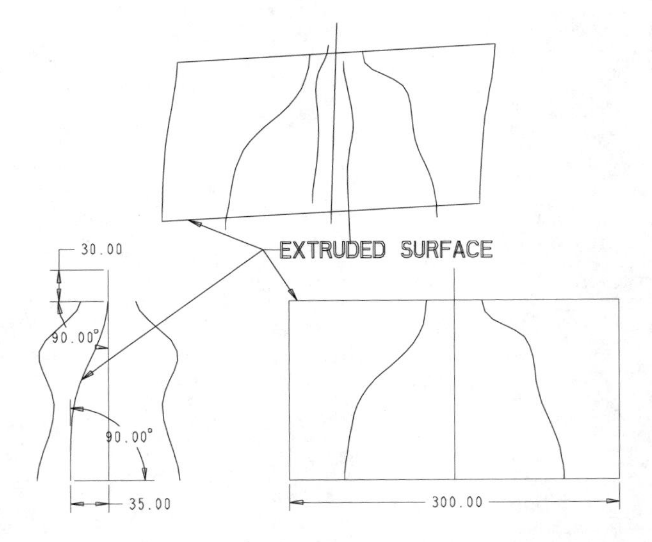

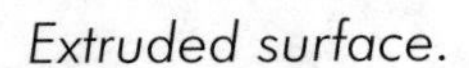
Extruded surface.

Once the surface is complete, the projected curve will be created. As with other features in Pro/ENGINEER, you have many options for creating projected curves, such as sketching or selecting a curve for projection, projecting normal to existing planar surfaces or datum planes, or creating a plane on the fly. The goal is to create a non-planar flexible part with features linked together according to predictable, defined objectives.

Consequently, the projected curve in this example will be created using the fifth existing curve and projecting it normal to the center datum plane of the part. The projected curve will then be copied and mirrored dependent, which creates the desired parent/child relationship.

> ✔ **NOTE:** *Another way to achieve a result similar to a projected curve to a surface is to use the "2 PROJECTIONS" datum curve option.*

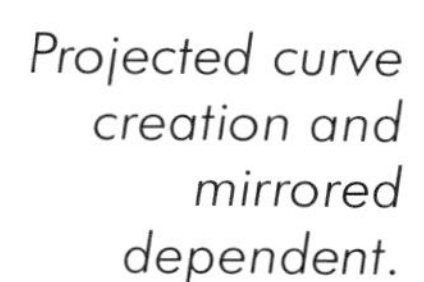
Projected curve creation and mirrored dependent.

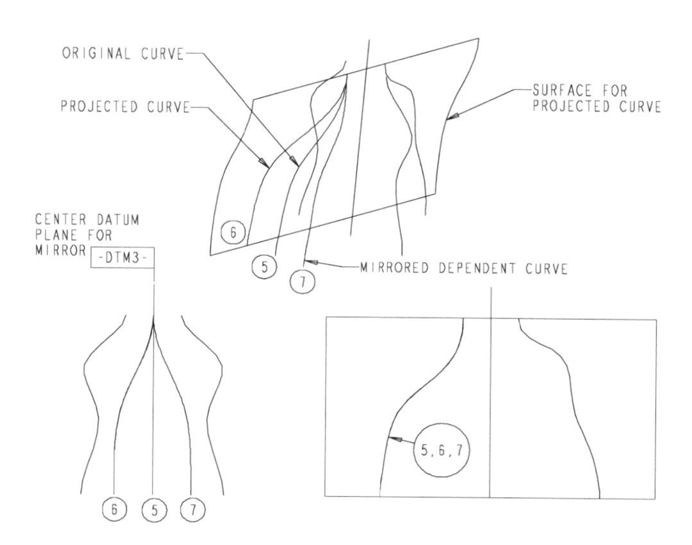

Major Change in Design

Now that the core of the laundry detergent bottle development has been revised, the big change is next. The original variable section sweep will be changed to correspond to the new projected curves. We will also take advantage of the wide range of tools available within Redefine to accomplish this change, resulting in a more complex non-planar shape. To execute this major change in design, take the following steps:

1. Cancel the Insert mode and retrieve the original variable section sweep.

✘ **TIP:**: *Use Modify and select the variable section sweep to quickly obtain information about the created sweep.*

2. The major change to the variable section sweep involves redefining. First, the curves that the variable section is swept along should be redefined. Second, the original cross section that is swept will be modified. All of these changes can be accomplished by using Redefine. Our blueprint has established the core for an effective and flexible non-planar part. When redefining a variable section sweep, you have

the ability to modify, redo, add, or show any of the curves used in creation of the sweep. Redefine the variable section sweep. Use Redo to replace the fifth curve with the projected curve.

3. Add an additional curve to the sweep by choosing Add. Select the mirrored dependent curve.

Redefine Redo and Add.

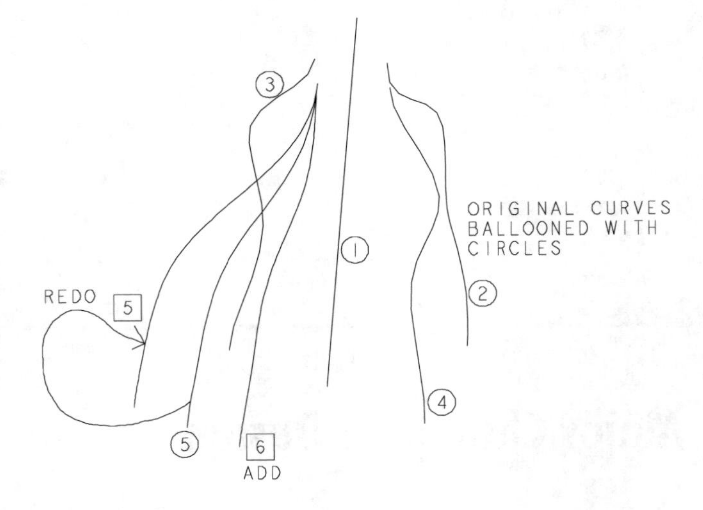

✔ **NOTE:** *Whenever a curve is added or removed during redefining of a variable section sweep, you must then regenerate the cross section to be swept. A quick way to add or remove more than one curve would be to sketch simple straight lines or circles as the cross section to be swept and then regenerate it until you are finished with adding or removing curves.*

4. Next, we will create a new cross section to be swept. The change to the original cross section is somewhat more complex because manufacturing requires a round top. To accommodate the round top, the new cross section to be swept will be created using conics forcing the rho value of the conic to develop into a circle.

Conic which can be used to define a circle.

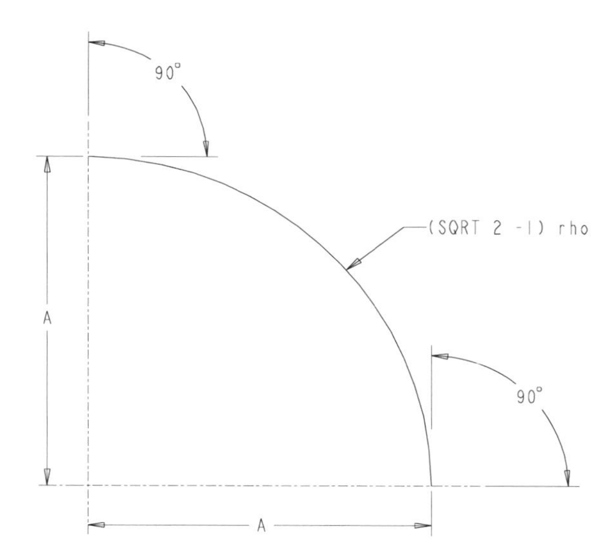

5. The new cross section will also use a two-point spline with tangency at both ends to create flexibility for any amount of curvature on the side.

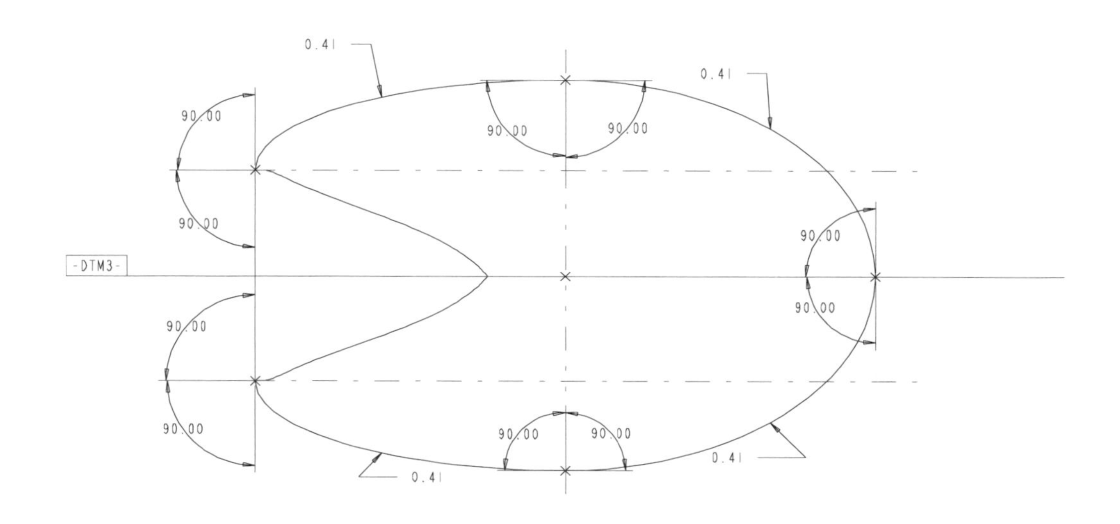

Redefined cross section to be swept.

Additional Control Through Graph Features and Relations

After the Redefine is complete, you will notice that with the variable section sweep the two-point spline used in the cross section does not give enough curvature to the side of the part. Recall that curvature is required by manufacturing in the blueprint. Steps for adding curvature follow.

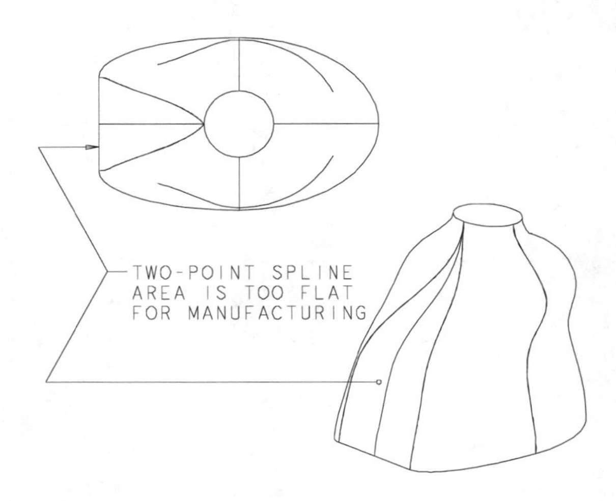

Laundry detergent bottle: more curvature needed.

---

1. Modify the tangency values for the conic from 90 degrees to 120 degrees.
2. The value for the tangency of the two-point spline at both ends should be added to the tangency values at the end of the conic to equal 180 degrees. This would yield a smooth intersection where the segments come together. However, the tangency value at the end of the conic must be 90 degrees at the top to generate the necessary circular bottle top. Simply put, the cross section to be swept must have different values at the bottom and the top.
3. A Graph feature will be added to gain additional control. The Graph feature, inserted just before the variable section sweep, is used to vary the value of the tangency as the protrusion is swept. Be sure to give the graph feature a logical name for easy identification later.

4. For the detergent bottle, one of the conic tangency values will be controlled and Sketcher relationships will be created to control the other tangency values.

Tangency values needed for sweep.

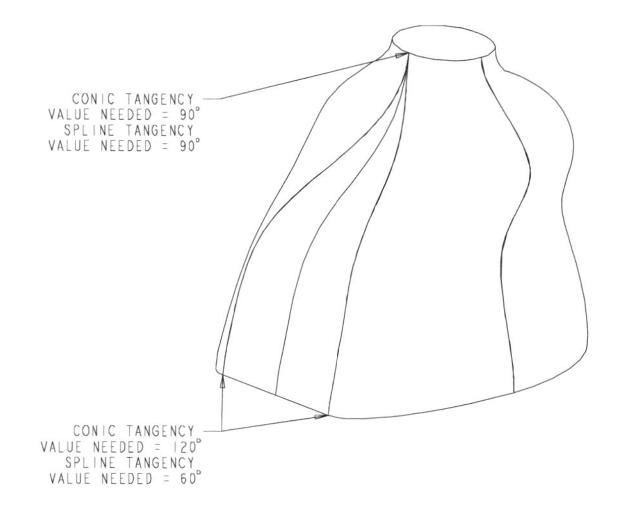

5. The graph feature will be a simple line supplying values from 120 to 90 degrees.

✔ **NOTE:** *A Graph feature uses the variable section sweep's length calculated from a value of zero at its starting point to a value of one at its ending point. The length of the graph will be extended to a value of 100 to simplify the creation of the graph as it is being sketched. Do not forget to then multiply the relationship used in the variable section sweep by the same factor (100).*

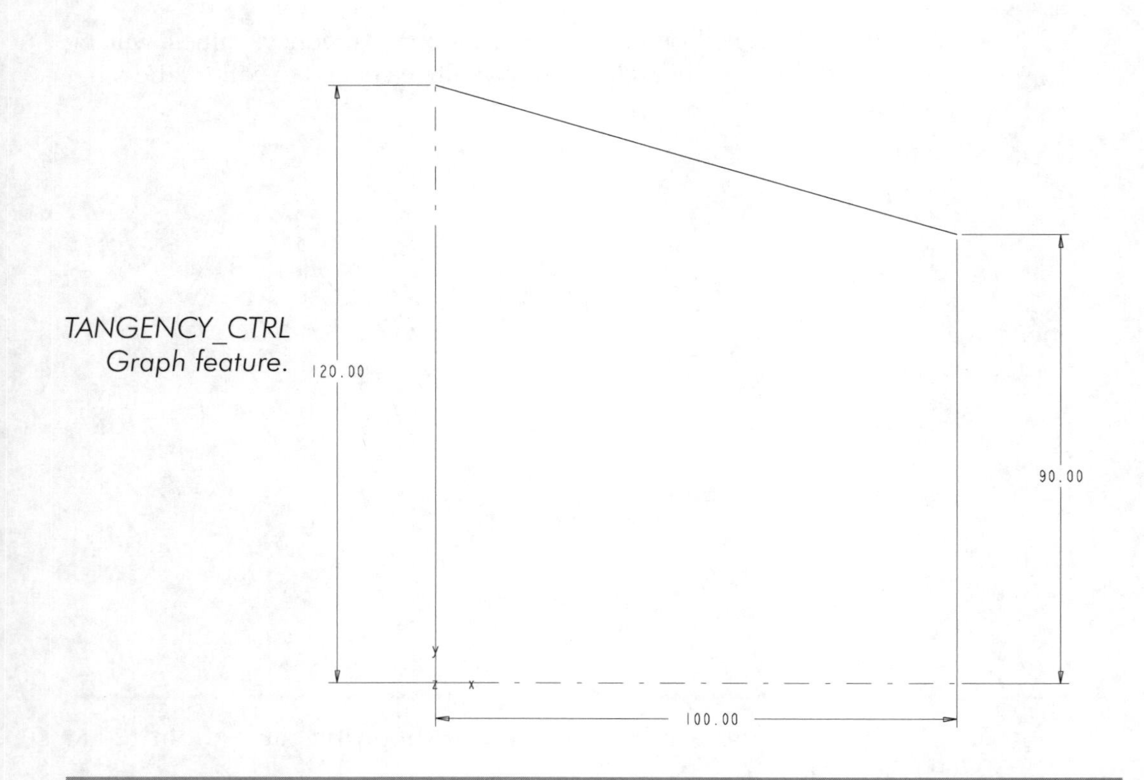

TANGENCY_CTRL Graph feature.

6. After the Graph feature is complete, cancel the Insert mode and begin redefining the variable section sweep.

7. The cross section (in Sketcher mode) to be swept will be the section to which the relationships will be applied. The relationship values are derived directly from the graph-generated values.

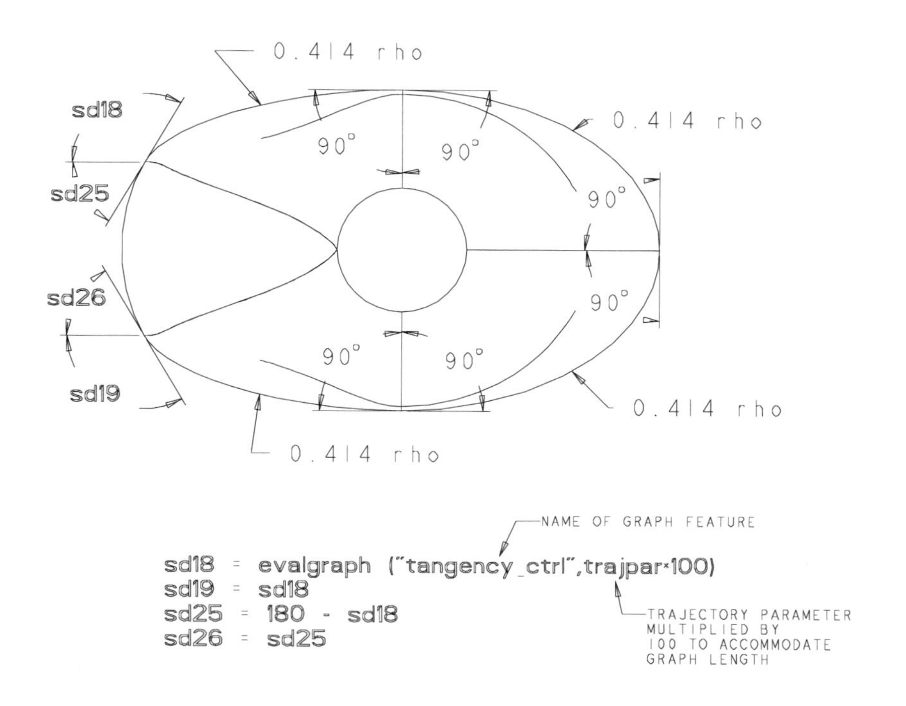

Relationships for cross section of variable section sweep.

8. Regenerate the cross section and the values will be updated. Complete the Redefine by selecting Done, and watch the new variable section develop the basic shape for the laundry detergent bottle. Note that the top is now circular because of the conic definition.

Completed variable section sweep.

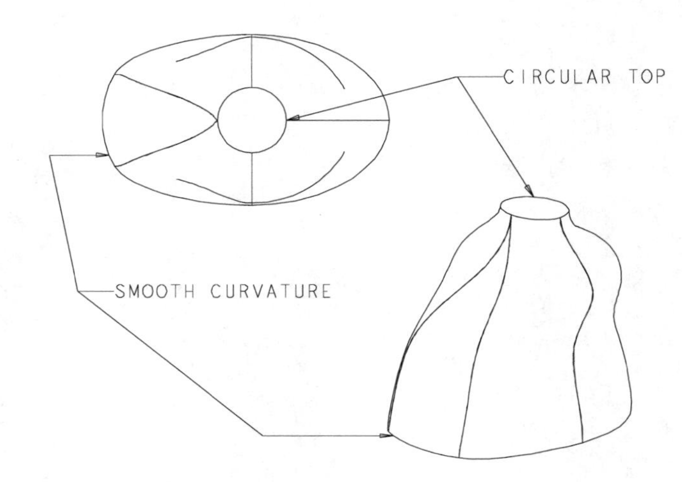

9. Add the material to the top of the part as a revolved protrusion to accommodate the circular screw-on cap. Use center lines and alignments in the Sketcher to create parent/child relationships to control the overall height.

Revolved top definition.

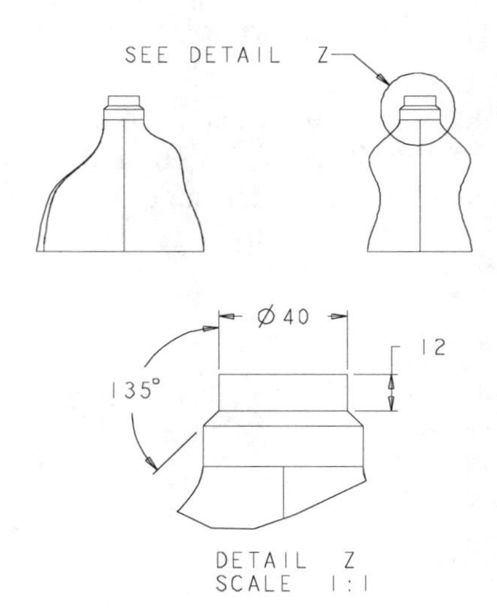

Adding A Handle

Like the body of the laundry detergent bottle, the handle will be created with non-planar shapes. Our objective is to create a handle with finger grips using a variable section sweep. In contrast to the body, the handle will be created using two curves to sweep along instead of six. Next, the handle will incorporate a unique cross section to vary the shape.

1. The first curve will be a conic to ensure smooth curvature as the handle is swept.
2. The second curve, which also serves as a guide for the finger slots, will be a spline. The guide allows for easy handling from the supermarket to home, another marketing requirement.

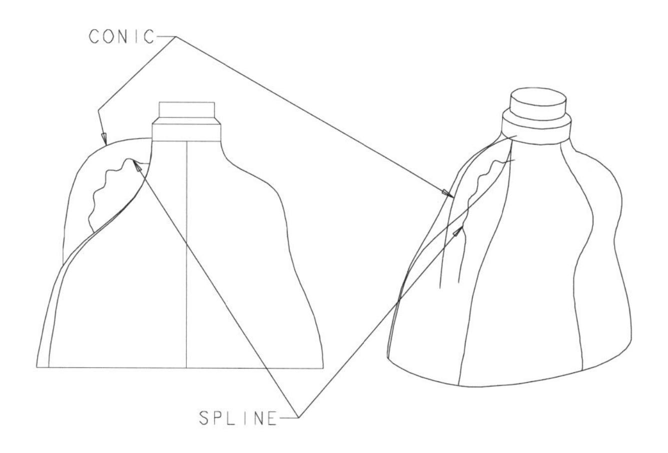

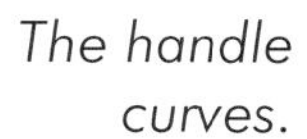
The handle curves.

---

3. The creation of the variable section sweep for the handle will have a unique cross section. The cross section follows the two curves, thereby maintaining a true radius along the first curve, and the shape varies along the finger curve.

- ❑ Develop the sketch for the cross section using an arc and a three-point spline with tangency at both ends.

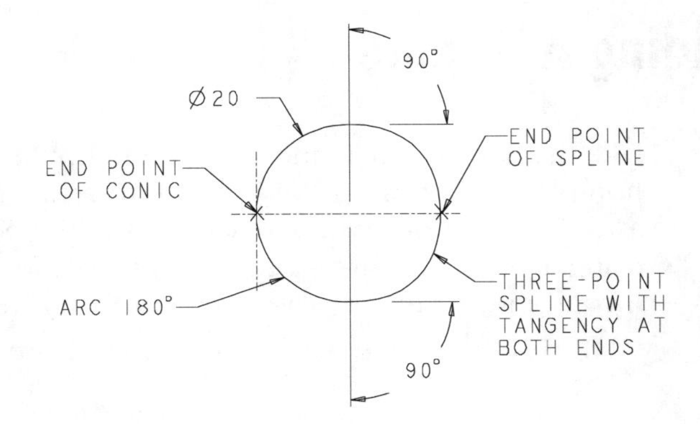

The cross section for the handle.

Rugged Appearance

One of the marketing department's major requirements is a detergent bottle which appears to be rugged. Although the completion of the handle contributed to this objective, we will enhance the rugged aspect of the bottle by adding a series of bumps or waves to the bottom of the part. For this purpose, we will use Pro/ENGINEER's Evaluate feature in conjunction with a curve created by equation.

The Evaluate feature, a recent addition to Pro/ENGINEER's base functions, allows for parameter definition to be based on existing features. Distances, angles, curve edge length, minimum radius, area, and even assembly clearances can be used to create a parameter for use later in product development. The Evaluate feature will be used to create parameters for use in creating a datum curve by equation.

1. Before creating the Evaluate feature, we will create two datum points by using the curve and surface option.

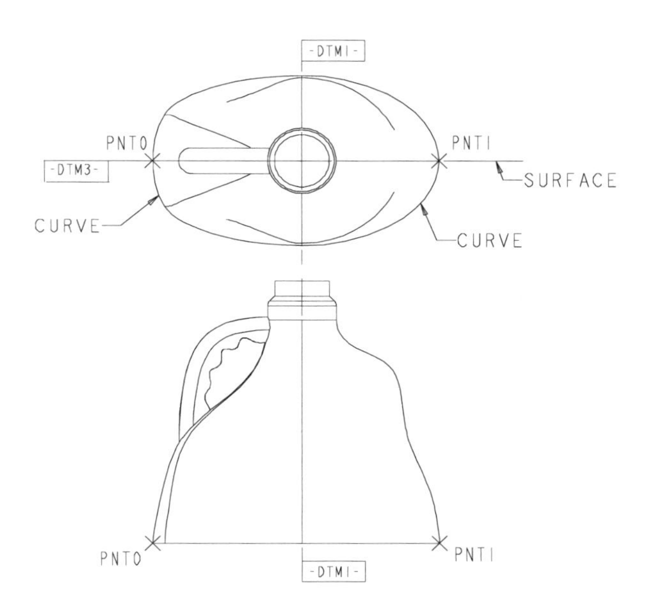

Datum point creation.

2. Next, two Evaluate features will be created. The first will represent the total width of the laundry detergent bottle, and the second, the width from the point on the handle side to the center of the part (datum 1).

✔ **NOTE:** *When creating an Evaluate feature, be sure to give it a logical name.*

TOTAL WIDTH

Measurement name	Measurement type	Current value
T_W	Distance	208.743382

Evaluate features.

WIDTH HANDLE SIDE

Measurement name	Measurement type	Current value
W_H_S	Distance	108.743382

3. We will now use the Evaluate features to aid in the development of a curve by equation. The curve by equation will create a series of bumps or waves on the bottom of the detergent bottle. A simple sine wave equation will be used to create this set of bumps or waves. The equation is to be shifted slightly to one side to equally space the bumps along the bottom of the bottle. In addition, the curve will be shifted upward to force the lowest point of the sine wave to the exact bottom of the part.

4. To create a curve by equation, we need a coordinate system. We will use the default coordinate system.

5. When creating a curve by equation, Pro/ENGINEER supplies an example equation that can be used to create a simple curve. The curve for the bottle is more advanced in that it incorporates Evaluate features during creation. As you input values for X,Y, and Z, test the result by selecting Show from the menu to view how the curve is being developed. (The curve may require a View Repaint.)

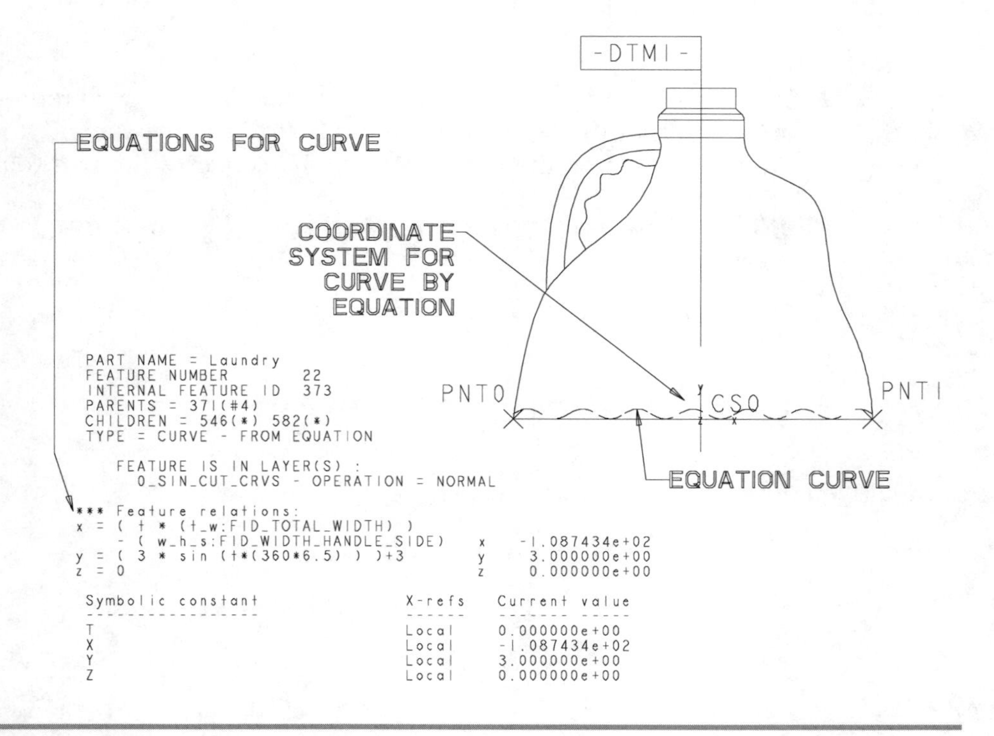

Coordinate system and equation.

Based on the equation, the sine wave will be shifted to the outermost edge on the handle side (X). The wave will consist of 6-1/2 complete sine waves, and will start at three units off the bottom (Y). After the curve is complete, create a cut in both directions using the edge within the Sketcher to select the equation-created datum curve.

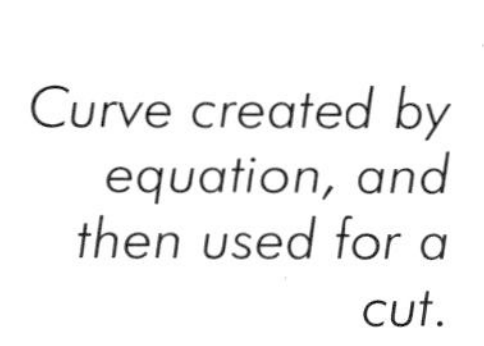

Curve created by equation, and then used for a cut.

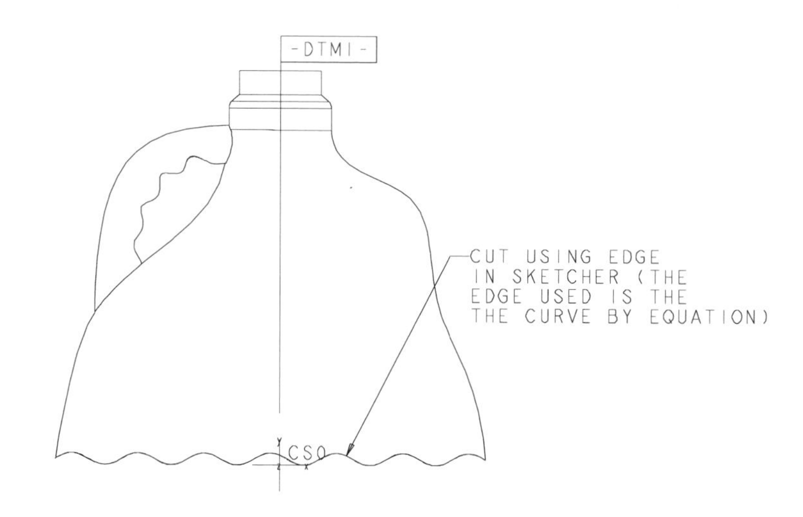

Final Details

To complete the laundry detergent bottle, a few more features will be added to the blueprint.

- ❑ Rounds will be added to ensure easy molding, and to enhance appearance.

Edge rounds added.

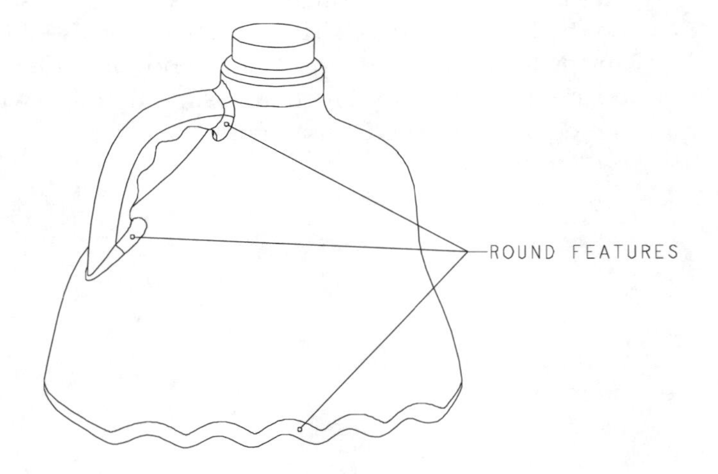

- Next is the Shell feature. The part will be shelled with constant wall thickness on all surfaces of the bottle with an opening at the top.
- The final items to add are the definition for the threads at the top of the part and a few rounds on the threads. Creation of the threads can be approached in one of two ways: (1) create cosmetic threads, and (2) create a helical protrusion on the part. In this case, the manufacturing department wants the complete part, including threads, to allow for rapid prototyping. Therefore, the second approach will be used to create the threads.

✔ **NOTE:** *Creating helical threads can be accomplished by using the advanced protrusion/cut option called Helical Sweep. The helical sweep is discussed in detail in Chapter 9.*

Completed laundry detergent bottle.

Summary

The detergent bottle example demonstrates how a part based on non-planar styles and functions can be developed. Going beyond basic cuts, protrusions, blends, and sweeps allows you to proceed to higher levels of design. In conjunction with variable section sweeps, using datum curves to develop the core of your blueprint helps you to create simple but powerful geometry. Note that simple does not mean that the geometry is not complex. The many non-planar shapes within the laundry detergent bottle clearly demonstrate the advantages to this approach.

Feature Reduction When Creating Complex Shapes

Introduction

In the course of developing a complex product, Pro/ENGINEER users find that the number of features may become very large, thereby decreasing performance and productivity. Consequently, reducing the total number of features whenever possible is important. However, there is a fine line between reducing the number of features and realizing your product design blueprint. In this chapter we will use the development of a fan to focus on a powerful technique for feature reduction. The technique is derived from the use of surfaces and shows how to reduce the number of features without losing design intent. In addition to the surface method, other examples are provided that help to reduce the number of features and improve regeneration time.

Fan Design Requirements

Completed fan.

The fan shown above will be used to demonstrate techniques to reduce features. Before the actual feature reduction occurs, the design blueprint or intent must be established. Be very careful during product development to avoid sacrificing design goals or requirements for feature reduction. The design requirements that establish the blueprint for developing the fan appear below.

- ❑ Assembly requirement: The total package area for the fan must not exceed 450mm in diameter and 100mm in height. This requirement will ensure that the completed fan will fit into existing housings.
- ❑ Assembly requirement: A mounting hole is to be created at the center of the fan. This hole must have the flexibility to be changed at any time to accommodate different mounting sizes.

- ❑ Engineering requirement: The number of blades must be changeable to accommodate different systems or to improve output.
- ❑ Manufacturing requirement: Rounds must be added to the interior portion of the hub to allow for easy molding.

With the design requirements established, we can begin building the reduced feature part.

Fan Construction

After reviewing the fan requirements, the product is broken down into three sets of features which describe the completed fan.

1. The hub or center portion of the fan.
2. Blades.
3. Minor details such as the mounting hole, rounds, and a keyway slot.

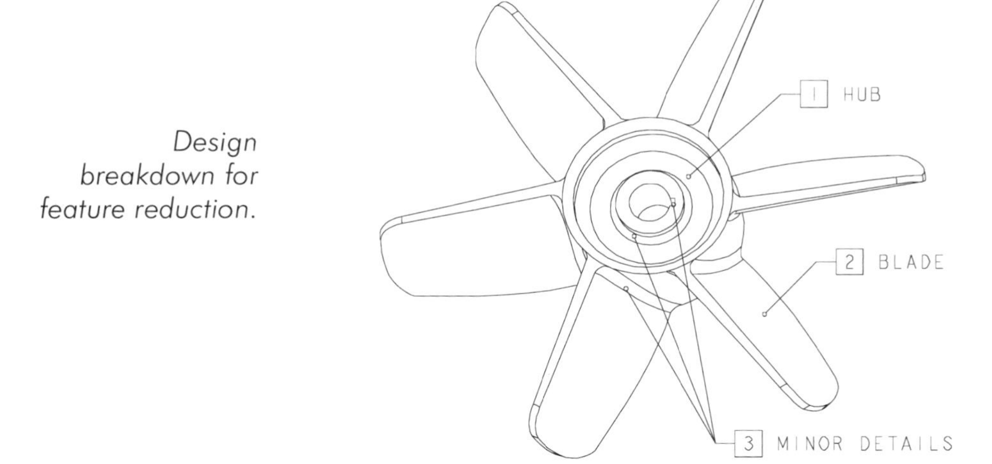

Design breakdown for feature reduction.

The three sets of features break down the design and establish a systematic way of approaching the design problem. We will begin with the hub by using basic methods for reducing features that are occasionally overlooked. Blade design (the following section) contains the most features and thus lends itself to the use of the surface technique for reducing feature numbers. Subsequent sections focus on other techniques which contribute to reducing feature numbers and regeneration time.

Hub Design

The hub of the fan is created with the use of a revolved protrusion. The sketch for the hub is designed to fulfill the height requirement and to reduce the total number of features for this portion of the design. Our approach for reducing feature numbers was to add additional entities to the sketch, rather than adding multiple revolved protrusions. Adding more entities to a sketch is a technique for which experience is the key ingredient. However, the following tips help in choosing whether to add additional entities to the sketch.

- Can the sketch be developed a portion at a time? If it can, create the small portion, Regenerate, and then select Done. Continue the development of the sketch by using **Redefine → Sketch**, add the additional entities to the section, and repeat until complete.

✘ **TIP:** *Using Redefine after completing portions of a sketch allows you to fall back to the finished portion of the sketch if you run into problems with sketch regeneration.*

- Is the sketch symmetrical? If it is, complete a half of the sketch and Regenerate. Once the sketch has regenerated, Mirror the sketch in the Sketcher.

Reducing features by adding sketcher entities a portion at a time.

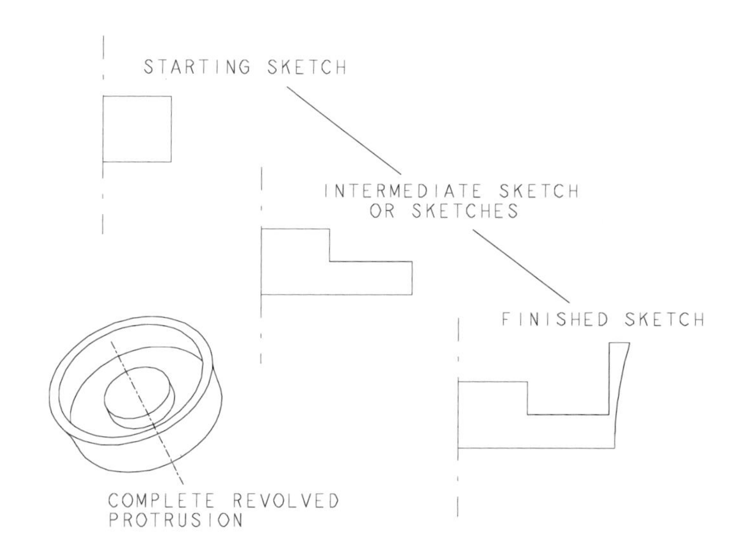

Reducing features by adding sketcher entities and mirroring in the Sketcher.

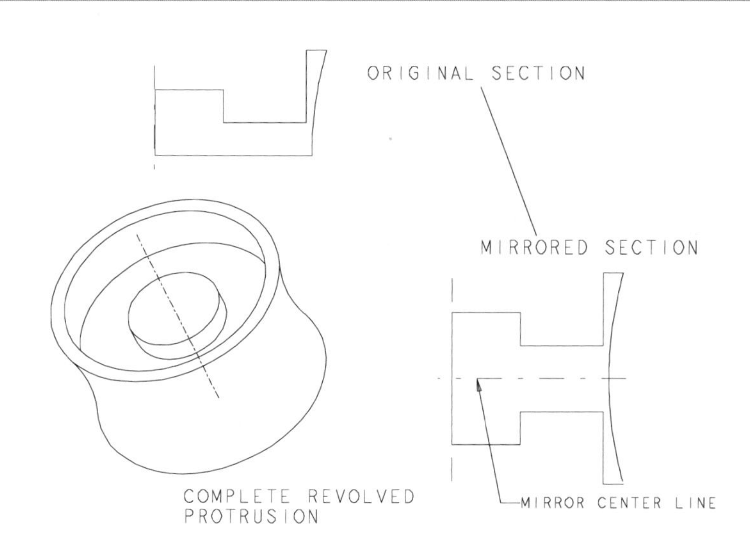

Both of the above techniques can be used to develop the fan hub. Later in the development of the fan, rounds will be added to the hub. The rounds could also be added to the original cross section as **Arc → Fillets** within the Sketcher. In this situation, experience comes into play. Adding the arcs to the section may make dimensioning and regeneration more difficult.

✔ **NOTE:** *As a general rule try to avoid adding* **Arc → Fillets** *to complex sketcher sections. Tangency conditions created by the arcs will more often than not make the sketch harder to work with and regenerate. Instead, add round features later in the part development.*

Blade Design

The blades will be created to fulfill two of the blueprint design requirements: outside diameter and flexible number of blades. The blade design contains many features to capture the design intent. In fact, over 15 features are used to complete a single blade. If the design for the first blade is duplicated for each additional blade, the total number of features will rapidly escalate. The techniques discussed below reduce the number of features for additional blades from over 15 to two.

Construction of the First Blade

The design for the first blade is quite complex, as it includes surfaces, datum planes, and surface merges. Construction steps are described below:

1. The first three features created establish the inside (edge of blade closest to the hub) and outside surfaces of the blade. The surfaces are revolved to form non-planar geometry. There are two inside surfaces so that a surface round can be created later in the development of the blade.

Inside and outside surfaces of the blade.

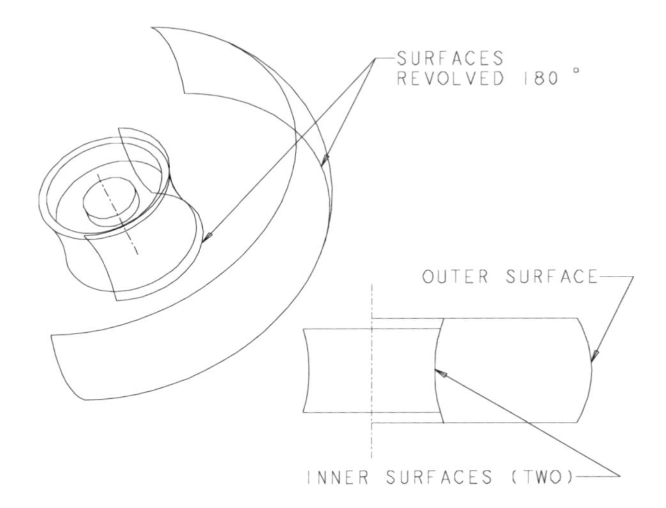

2. Following the inside and outside surface creation are two projected datum curves. In order to facilitate feature reduction, each projected curve contains two sketched curves. These curves will be used to define the width at the bottom and top of the blade. Note that the curves are projected onto the non-planar inner and outer surfaces created in step 1.

Projected curves for width of blade.

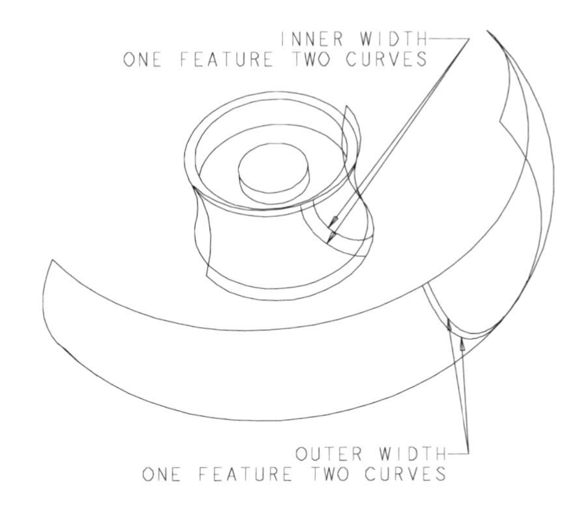

3. Next is creation of the sides of the blade. Use an advanced surface boundary command sequence for each side of the blade.

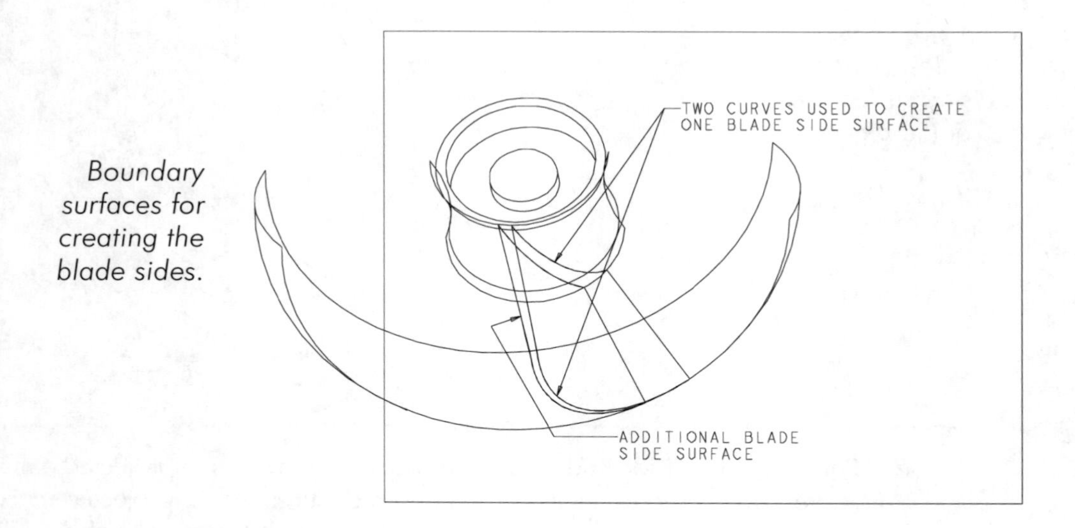

Boundary surfaces for creating the blade sides.

---

4. The next two features define and control the top and bottom surfaces of the blade. The two features are revolved surfaces with a dimension applied to control the angle from the outside edge of the hub to the outer surface of the blade.

Top and bottom surfaces of the blade.

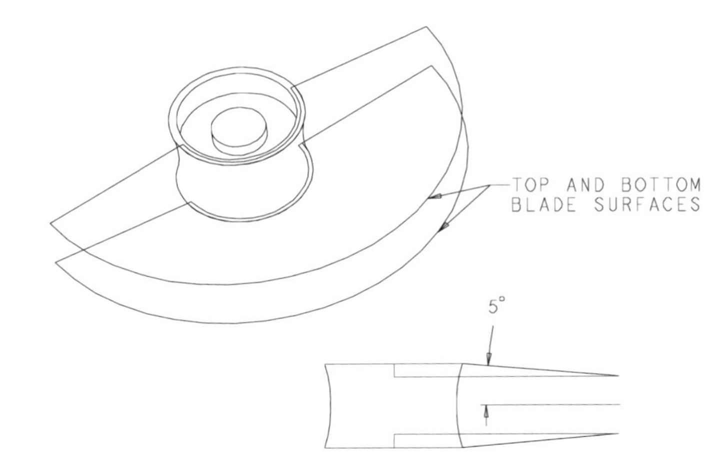

5. After completion of all surfaces which fully define the first blade, the **Surface → Merge** command sequence is used to combine them. In fact, the next two features merge the two inside surfaces with the two sides of the blade.

Inside surfaces merge with blade side surfaces.

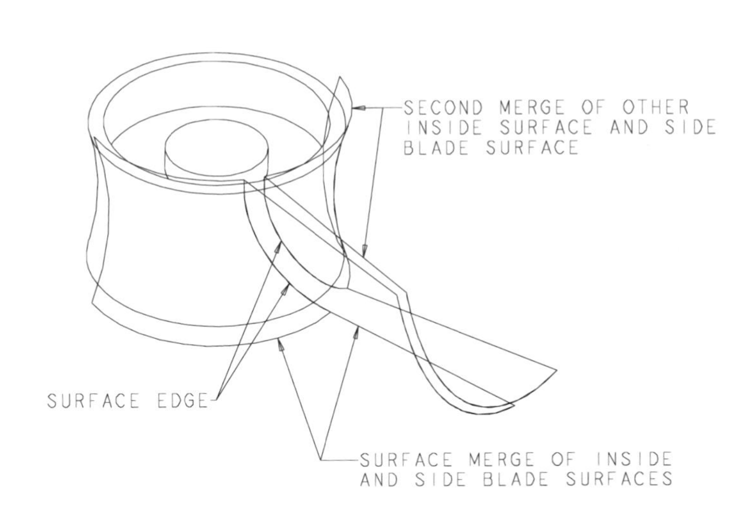

6. A round feature is added to the blade before the final **Surface → Merge** features are added. The round is developed using the surface edges created by the previous merges.

Rounds added to base of blade.

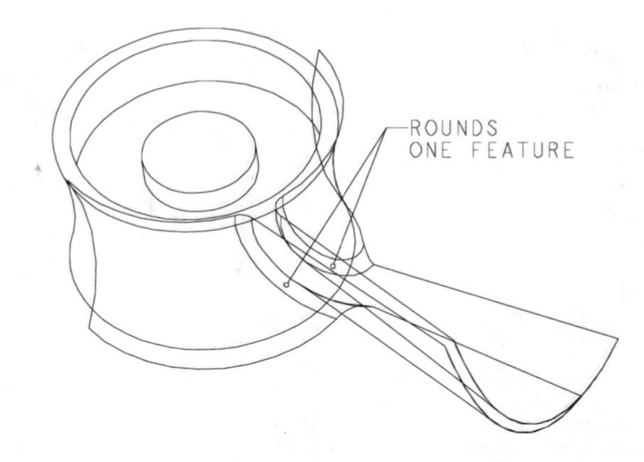

✔ **NOTE:** *The ability to add round features to surfaces is often overlooked. Round features added to surface edges (not solid edges) can add flexibility to a design.*

7. Four **Surface → Merge** features are created to complete the blade.

Final Surface → Merge features.

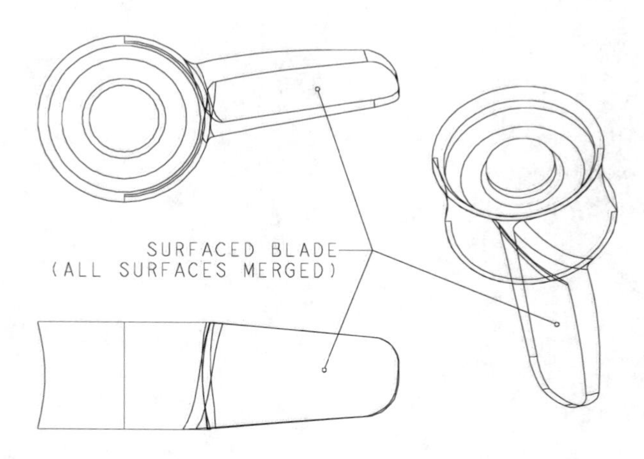

Adding More Blades

After the completion of the blade design the focus shifts to incorporating additional blades. The additional blades will not be created in the same manner. Instead, another surface technique is used to duplicate the original blade. Using **Surface → Transform → Rotate → Copy** allows duplication of the original blade without losing any of the original blade's design intent. The most important reason for this approach is feature reduction. The new surface is copied without the baggage of all additional features that were originally used to create it.

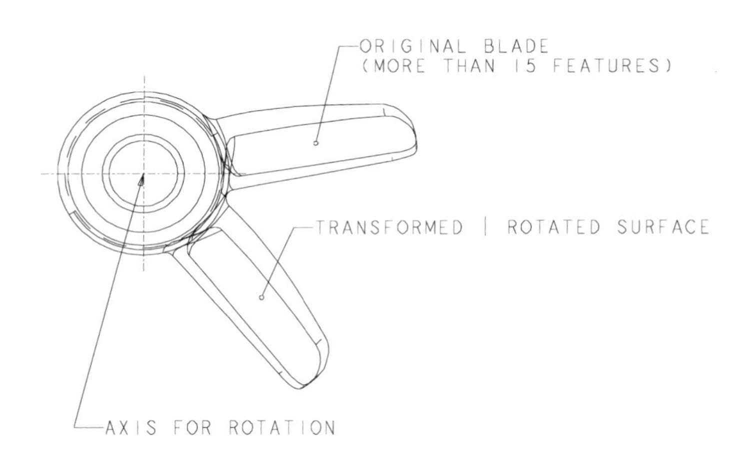

✔ **NOTE:** *In the* **Surface → Transform** *menu you have other options for duplicating surface geometry. In addition to the Rotate option shown above, consider the Translate (moving surface in X,Y,Z direction) and Mirror options.*

The surface can now be used to create a protrusion. Use the following command sequence:

Feature → Create → Protrusion → Use Surf

The transformed surface and the protrusion can be grouped together using a Local Group option and then patterned. The pattern for the blade will be driven by the value for the transformed rotated surface contained in the Local Group.

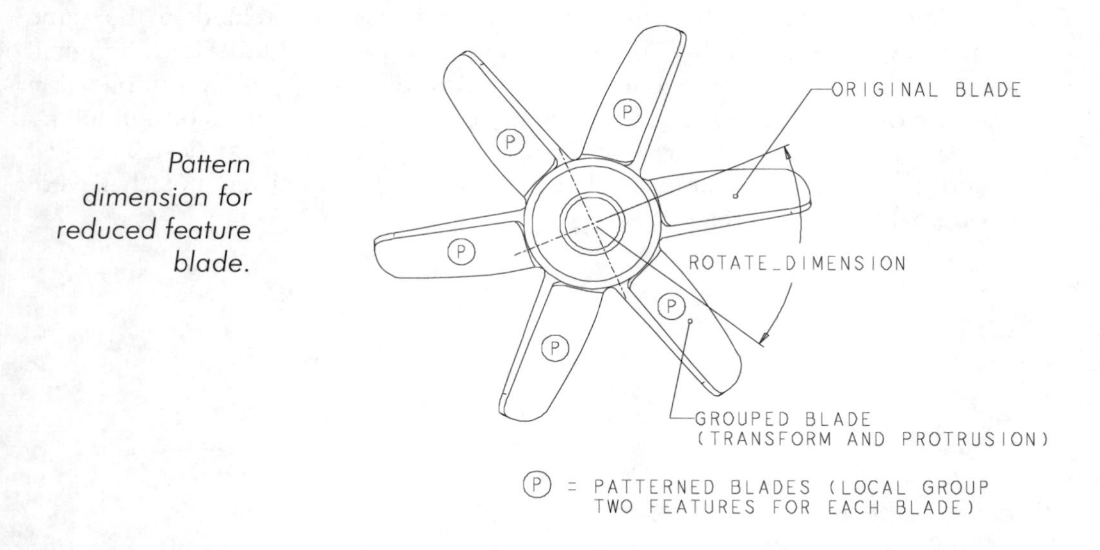

Pattern dimension for reduced feature blade.

The features that went into the development of the first blade are now reduced to two. Most importantly, this technique will allow you to pattern the design about the rotated dimension of the transformed surface. Any design intent established in the original is passed on to its children, the additional blades.

Other Feature and Regeneration Time Reduction Techniques

When the patterned blades are complete, only minor details remain. Although details are considered minor, they may be numerous. Another technique which helps reduce feature numbers is to create multiple identical pick and place features as a single feature. Multiple pick and place features such as rounds, chamfers, or drafts are especially amenable to combining into a single feature.

Multiple (Round) pick and place features created as a single feature.

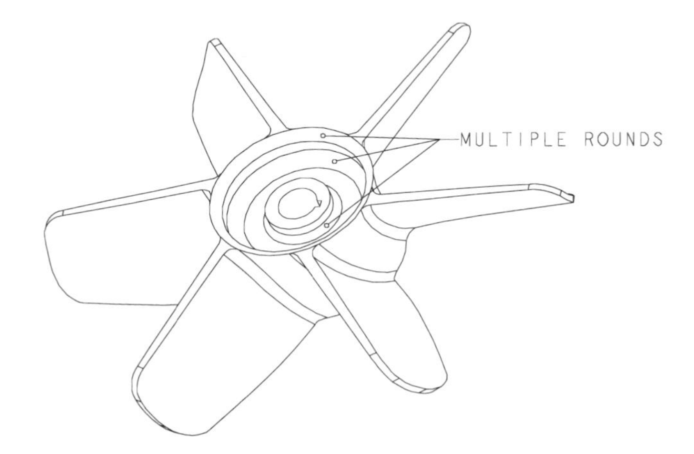

✔ **NOTE:** *When creating multiple identicalpick and place features as a singlefeature, thefollowing simple rules are recommended: (1) create thefeature in a logical area or portion of the part, and (2) do not go overboard.*

The last couple of features added to the fan part include the mounting hole feature. The mounting hole is important not because it reduces feature numbers, but because it will greatly improve regeneration or performance. According to the blueprint, this hole is subject to frequent changes. Because the hole is near the end of the design sequence, any modification to the hole will result in a very short regeneration time. In contrast, if the hole were created with or immediately after the hub, the blades would regenerate every time the hole changed, thereby increasing regeneration time.

Summary

Completed fan.

During product development, feature numbers can greatly affect regeneration time and productivity. Given that the product design blueprint is not compromised, feature reduction can boost performance. The use of the **Surface → Transform** technique enables the duplication of design intent without duplicating feature numbers. In addition, creating multiple identical pick and place features as a single feature contributes to reducing feature numbers. Because Pro/ENGINEER or regenerates sequentially, both the number and order of features are important.

Developing Design Options Using Pro/PROGRAM

Introduction

Playing the "what if" game can be exciting throughout product development. All types of users have experienced the benefits of using parametrics and relationships to control a design. Using parametrics and relationships with Pro/PROGRAM goes a step beyond the usual. In this chapter we will briefly discuss a common misconception regarding Pro/PROGRAM. Next, a simple problem using Pro/PROGRAM to modify a part is presented, followed by a complex problem illustrating the many benefits of Pro/PROGRAM.

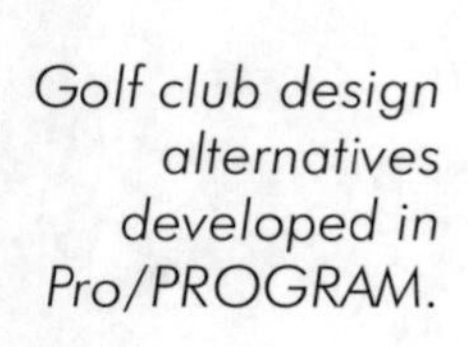
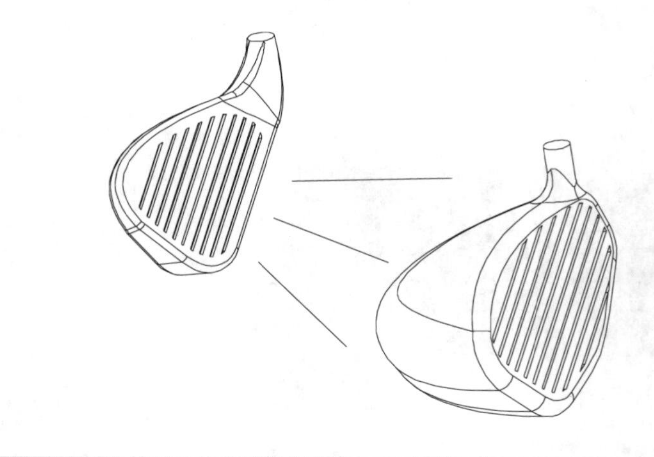

Golf club design alternatives developed in Pro/PROGRAM.

Misconceptions about Pro/PROGRAM

Pro/PROGRAM is an extremely powerful tool but many users misunderstand its function. Pro/PROGRAM allows you direct access to a model's feature construction sequence which in turn gives you the ability to directly edit the design. When you use input parameters and conditional statements (IF, ELSE), you obtain a higher level of control over all parameters and relationships developed in a model.

Pro/PROGRAM is not an application programmer's interface (API) in that it is a direct link to only the existing data created. An API, such as Pro/ENGINEER's Pro/DEVELOP, supplies a complete set of tools for programmers to link into Pro/ENGINEER's database. Advanced users should not be discouraged from diving into Pro/PROGRAM and experiencing its power.

Pro/PROGRAM consists of the following five basic areas:

- ❑ Header information containing version, revision number, and a listing for part or assembly.
- ❑ Input area for user-supplied parameters.
- ❑ Relations area for user-created relationships.
- ❑ Complete feature listing, even if a feature is suppressed.
- ❑ Mass property area used to update mass properties every time geometry changes.

All areas, except the header information area, can contain conditional statements.

Controlling a Cube Using Pro/PROGRAM

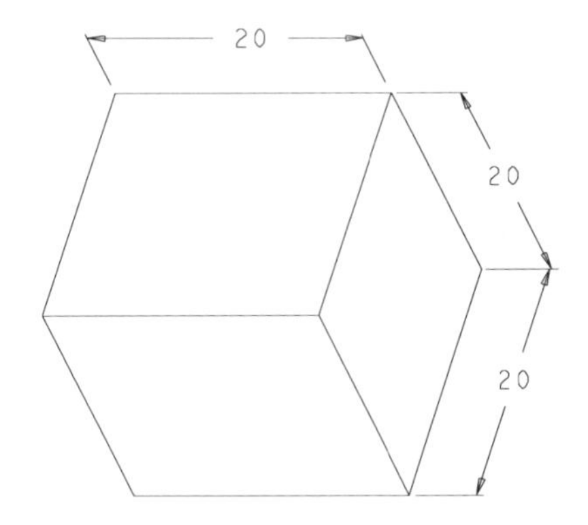

A single value cube.

The design blueprint for the cube specifies that a single value controls the size of the cube. During every regeneration of the part, you are prompted for the cube size.

The first feature is a protrusion sketched as a square and given height, width, and depth. After the protrusion is created, each dimension's symbolic value will be modified to correlate to height (H), width (W), and depth (D). This technique is accomplished by selecting **Modify → DimCosmetics → Symbol**, and then choosing the dimensional parameter you wish to change. The previous technique will help clean up your design and generally makes relationships and Pro/PROGRAM easier to use and implement. (D0 will be changed to equal H and so on.)

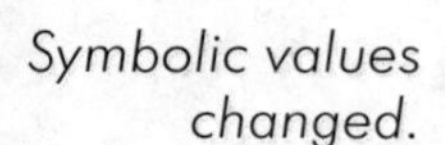

Symbolic values changed.

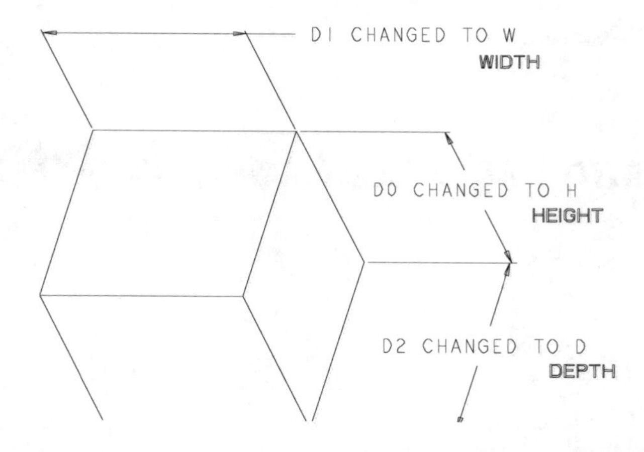

Since the part is to be a cube, relationships can be created to set the value for the width and depth equal to a single parameter, or height. A shortcut for adding relationships is to use the Modify command. By selecting Modify and choosing the width or the depth parameter, you can input a new value. Instead of inputting a new dimensional value, enter the symbolic value for the height (H). This action will automatically create the relation width or depth equal to height (W = H and D = H).

Relations created through Modify.

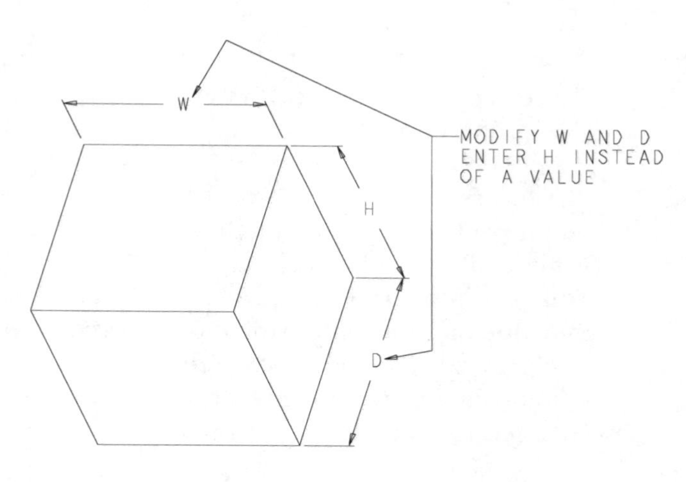

The final step for the cube design is to incorporate the Pro/PROGRAM. From the Part menu, select **Program → Edit Design**. This selection will activate a systems window containing information about the current model as shown below.

```
VERSION XX.0
REVNUM XX
LISTING FOR PART CUBE
INPUT
END INPUT
RELATIONS
      W = H
```

(Relationships may also be added directly during Edit Design.)

```
      D = H
END RELATIONS
ADD FEATURE (initial number 1)
INTERNAL FEATURE ID 1
TYPE = FIRST FEATURE
FORM = EXTRUDED
SECTION NAME = S2D0002
DEPTH = BLIND
FEATURE'S DIMENSIONS:
H = 20
D = 20
W = 20
END ADD
MASSPROP
END MASSPROP
```

✔ **NOTE:** *When you first enter Edit Design, the only design listing available pertains to the current model. After editing the design, you will be prompted to "incorporate the design changes into the model, Yes or No." Answering "No" will create a file called "modelname.als," or "...pls." The presence of this file will create an additional menu the next time Edit Design is selected. The next option would be to select either From Model or From File. Be aware that the design "from model" and "from file" may be quite different. From Model will reflect the current state of the model, while From File will reflect the instructions and features that were contained in the last edit.*

A line will be added in the input area to automatically prompt the user to input the value for the height (H) of the part. This input statement will be followed by an optional prompt explaining the purpose of the input value. Although the prompt for the input value is optional, it is the best technique. The incorporated program appears below.

```
VERSION XX.0
REVNUM XX
LISTING FOR PART CUBE
INPUT
      H NUMBER
      "Input the value for the height of the cube. Note:
      Width and Depth = Height"
END INPUT
RELATIONS
W = H
D = H
END RELATIONS
ADD FEATURE (initial number 1)
INTERNAL FEATURE ID 1
TYPE = FIRST FEATURE
FORM = EXTRUDED
SECTION NAME = S2D0002
DEPTH = BLIND
FEATURE'S DIMENSIONS:
H = 20
D = 20
W = 20
END ADD
MASSPROP
END MASSPROP
```

✔ **NOTE:** *Input statements for Pro/PROGRAM can contain any Pro/ENGINEER parameter (Number, String, or Yes_No). Notice that the height parameter is a number.*

Whenever a regeneration occurs, Pro/PROGRAM requires that you select Current Values, Enter Values, or Read File.

- ❑ Current Values: Pro/PROGRAM maintains all current parameter values used in the program.
- ❑ Enter Values: Activates another menu allowing you to select any appropriate parameter for modification (for the cube, H).

- Read File: Allows Pro/PROGRAM to directly read in input values rather than the user manually inputting values.

Expanding the Use of Pro/PROGRAM

To expand upon the use of Pro/PROGRAM, we will develop a golf club. A typical golfer will carry a golf bag containing 11 clubs, or four woods (1,3,5, and 7) and eight irons (3 through 10, 10 being a pitching wedge). Through a series of inputs and conditional statements in a Pro/PROGRAM, you can use a single club to generate any wood or iron golf club.

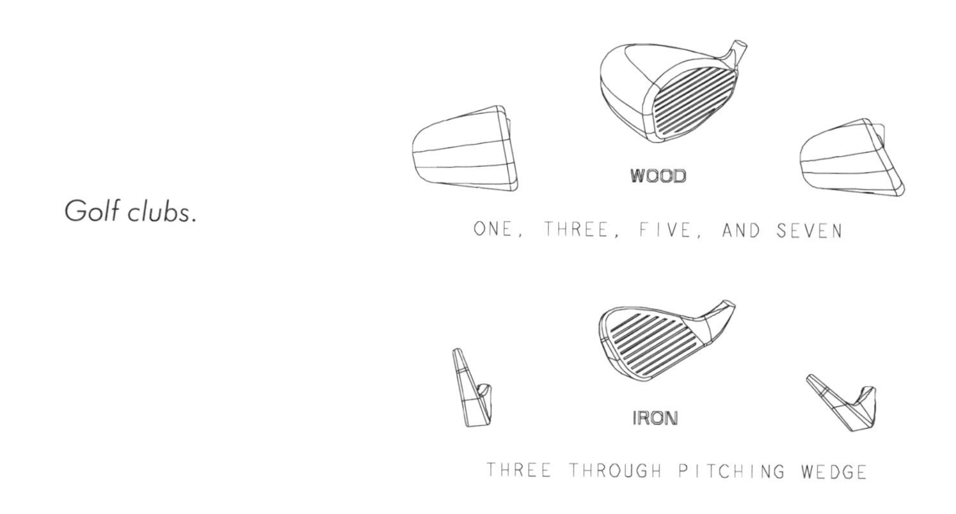

Golf clubs.

The Golf Club Blueprint

The main objective for the golf club design is for a single part to create any golf club. Assume that the sales and marketing departments have voiced a desire for a demonstration tool which can be used to show the golf club product line. The main requirements or inputs which establish the golf club blueprint include the following:

- Is the golf club a wood?
- What is the golf club's loft or number?
- If the golf club is an iron, does it have a cavity back?

Using Pro/PROGRAM to Resolve Design Requirements

Initially, the task of creating a part which can dramatically become another part with a completely different shape (iron to a wood) may seem to be an impossible task. Using Pro/PROGRAM can make this task a reality.

Is the Golf Club a Wood?

As you begin to break down the design the first item to focus on is whether the club is to be a wood. Three items within Pro/PROGRAM will be used to resolve this question.

1. The first function is to create a parameter which will be used as an input statement in Pro/PROGRAM. This parameter is similar to the one mentioned in the cube problem (H NUMBER), except that the parameter will be a Yes_No rather than a Number.
2. Create an input statement in Pro/PROGRAM.

✔ **NOTE:** *A parameter can be added two ways in Pro/ENGINEER. The first method is to select* **Setup → Parameter → Create** *and specify whether the parameter is to be a Number, String, or Yes_No. The specified parameter should then be added to the input section of Pro/PROGRAM. The second method is to select* **Program → Edit Design** *and directly apply the parameter to an input statement. If you choose to apply the parameter using the second method, be sure to specify the type of parameter. If you do not specify the parameter type, Pro/PROGRAM will assume the default, or a Number.*

3. A Pro/PROGRAM conditional statement can now be added by using Edit Design to determine whether the club is to be a wood or iron.

```
INPUT
        CLUB_WOOD YES_NO
END INPUT
RELATIONS
END RELATIONS
        IF CLUB_WOOD == NO
```

(Features used to create an iron are located here.)

```
        ELSE
```

(Features used to create a wood are located here.)

```
        END IF
```

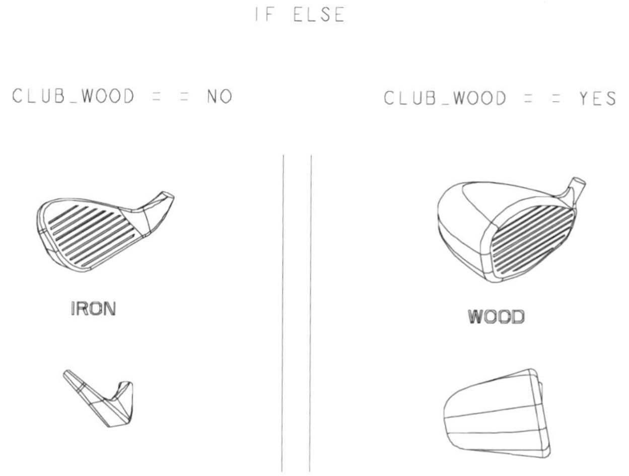

IF ELSE

CLUB_WOOD = = NO

CLUB_WOOD = = YES

IRON

WOOD

Pro/PROGRAM conditional statement.

✔ **NOTE:** *A conditional statement in the feature listing area will automatically force any feature not meeting the required condition to be suppressed.*

Before addressing the remaining blueprint requirements for Pro/PROGRAM—the loft or number of the club, and whether the club has a cavity back if it is an iron—, the part should be flexed. The key to flexing the golf club for Pro/PROGRAM is to focus on any other parameter or features that you wish to control. Flexing the part at this point will enable quick and easy implementation of the Pro/PROGRAM.

- For both the iron and the wood, the geometry must be built in to change the club's loft or angle. The golf club is constructed so that a single dimension can be modified to make any loft change. The symbolic value for the dimension is modified to a logical name for clarification, that is, IRON_LOFT and WOOD_LOFT. This dimension should be modified and regenerated at each loft position to satisfy the design requirements.

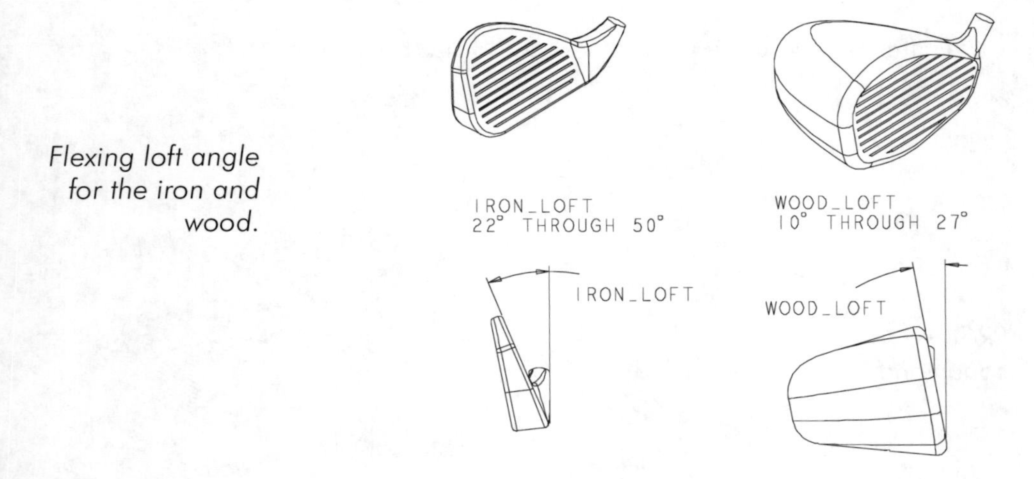

Flexing loft angle for the iron and wood.

During flexing of the iron a design issue becomes apparent. The changes to the iron loft illustrate that the sole of the club requires additional control depending on the value of the loft angle. The sole angle will require a unique value relative to the loft angle. The symbolic value for the sole angle will be changed to IRON_BASE_ANGLE.

Additional iron control.

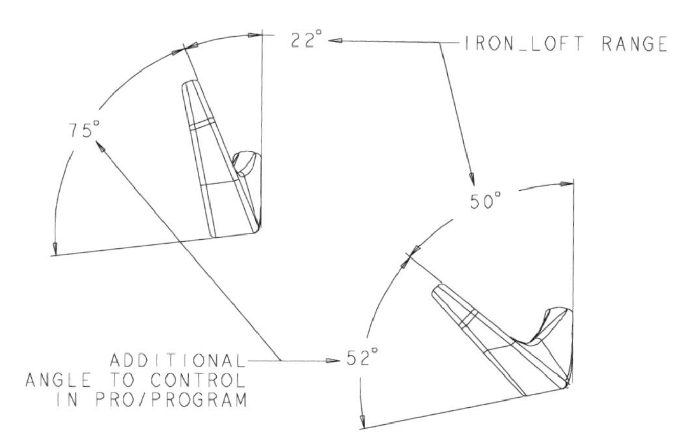

- The iron will have a set of features to be used for creating a cavity back. The features include a Datum curve (defining the profile for the cavity back), a Cut (using the profile from the curve), and two Rounds (smoothing out the edges created by the cut). This set of features should be tested to verify that the cavity back can be suppressed or resumed with each new loft value.

Flexing the cavity back feature for the iron.

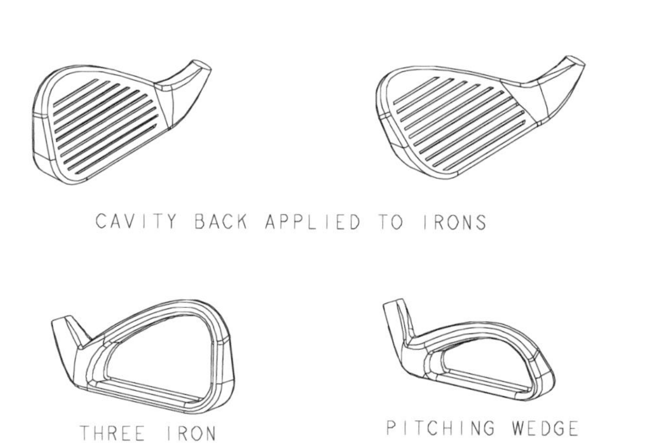

Golf Club Loft or Number

To capture the golf club's loft in Pro/PROGRAM a series of conditional statements and relations will be created. Before implementing the loft in Pro/PROGRAM, we will briefly describe loft. A golf club's loft is directly related to the club number. For example, the loft angle for a five iron is 28 degrees, and for a five wood, 21 degrees.

To apply the golf club's number to the Pro/PROGRAM, an additional parameter will be added. The parameter will be added directly along with all relationships and conditional statements by using **Program → Edit Design**.

```
INPUT
CLUB_WOOD YES_NO
        IF CLUB_WOOD == NO
            CLUB_NUMBER NUMBER (parameter used to relate
        club number to loft)
        "ENTER CLUB NUMBER FOR IRON 3-10"
        END IF
        IF CLUB_WOOD == YES
        CLUB_NUMBER NUMBER
        "ENTER CLUB NUMBER FOR WOOD 1,3,5,7"
        END IF
END INPUT
RELATIONS
        IF CLUB_NUMBER == 1
        WOOD_LOFT = 10
        END IF
        IF CLUB_NUMBER == 3
        IRON_LOFT = 22
        IRON_BASE_ANGLE = 75
        WOOD_LOFT = 15
        END IF
        IF CLUB_NUMBER == 4
        IRON_LOFT= 25
        IRON_BASE_ANGLE = 72
        END IF
        IF CLUB_NUMBER == 5
        IRON_LOFT= 28
        IRON_BASE_ANGLE = 70
        WOOD_LOFT = 21
        END IF
        IF CLUB_NUMBER == 6
```

```
          IRON_LOFT= 32
          IRON_BASE_ANGLE = 68
          END IF
          IF CLUB_NUMBER == 7
          IRON_LOFT = 36
          IRON_BASE_ANGLE = 64
          WOOD_LOFT = 27
          END IF
          IF CLUB_NUMBER == 8
          IRON_LOFT = 40
          IRON_BASE_ANGLE = 60
          END IF
          IF CLUB_NUMBER == 9
          IRON_LOFT = 45
          IRON_BASE_ANGLE = 56
          IF CLUB_NUMBER == 10
          IRON_LOFT = 50
          IRON_BASE_ANGLE = 52
          END IF
    END RELATIONS
          IF CLUB_WOOD == NO
(Features used to create an iron are located here.)
          ELSE
(Features used to create a wood are located here.)
          END IF
```

Selecting Enter Values after the edit is complete will activate another menu allowing you to select appropriate parameters for modification (CLUB_WOOD YES_NO or CLUB_NUMBER NUMBER).

✔ **NOTE:** *Notice in the input section that a conditional statement was added to force different prompts depending on whether the golf club is wood or iron.*

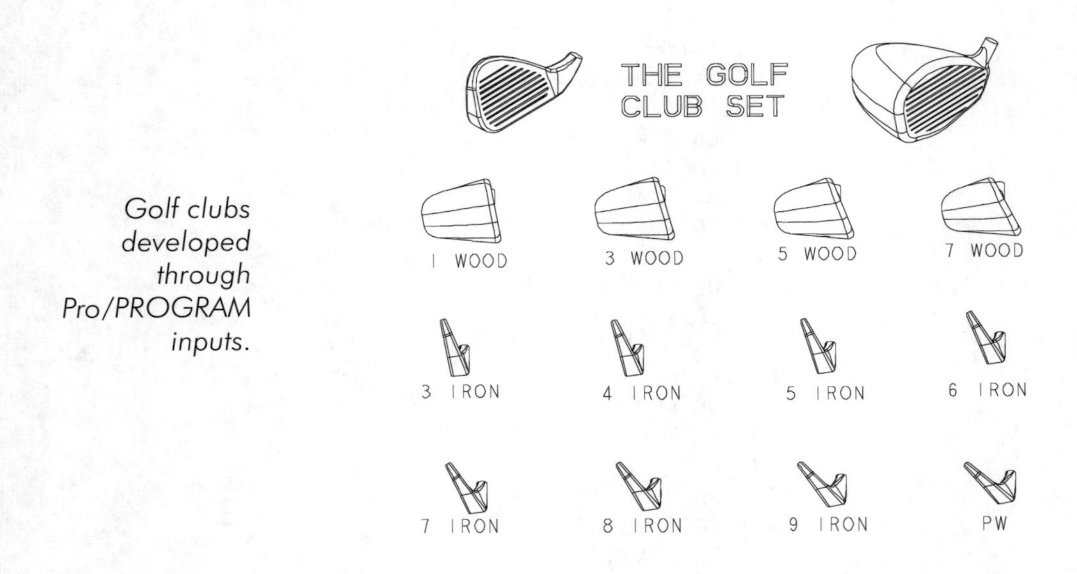

Golf clubs developed through Pro/PROGRAM inputs.

Deciding on Cavity Backs for Irons

The final portion of the design blueprint is related only to irons. During the flexing of the part the set of features which define the cavity back could be easily suppressed or resumed at any loft angle. This ability will allow for easy implementation of either non-cavity or cavity back irons in Pro/PROGRAM.

Up to this point most of the conditional statements applied to Pro/PROGRAM have been applied to the input and relations area of Pro/PROGRAM. The only exception is the case of defining whether a club is wood or iron, which is accomplished in the features listing area.

```
IF CLUB_WOOD == NO
```
(Features used to create an iron are located here.)
```
ELSE
```
(Features used to create a wood are located here.)
```
END IF
```

To add the option for a cavity back on the irons, an additional input parameter (CAVITY_BACK YES_NO) and an additional conditional statement in the irons feature area must be added. These edits to the program will add another prompt for the user. When adding the conditional statement to the iron feature area, make a note of the features that create the cavity back. The extra bit of caution will make the features easier to find during Edit Design, and will help to prevent accidental suppression of features not related to the cavity back.

```
INPUT
CLUB_WOOD YES_NO
"IS THE GOLF CLUB A WOOD?"
IF CLUB_WOOD == NO
CLUB_NUMBER NUMBER
"ENTER CLUB NUMBER FOR IRON 3-10"
      CAVITY_BACK YES_NO
      "DO YOU WANT A CAVITY BACK?"
END IF
IF CLUB_WOOD == YES
CLUB_NUMBER NUMBER
"ENTER CLUB NUMBER FOR WOOD 1,3,5,7"
END IF
END INPUT
RELATIONS
```
(ALL PREVIOUS RELATIONS)
```
END RELATIONS
IF CLUB_WOOD == NO
```
(Features used to create an iron are located here.)
```
      IF CAVITY_BACK == YES
```
(Features used to create the cavity back to the iron are located here.)
```
      END IF
ELSE
```
(Features used to create a wood are located here.)
```
END IF
```

Selecting Enter Values after the edit is complete will again activate another menu allowing the user to select an appropriate parameter for modification (CLUB_WOOD YES_NO, CLUB_NUMBER NUMBER, and if an iron CAVITY_BACK YES_NO). With the cavity back option placed in the program, all design requirements are complete.

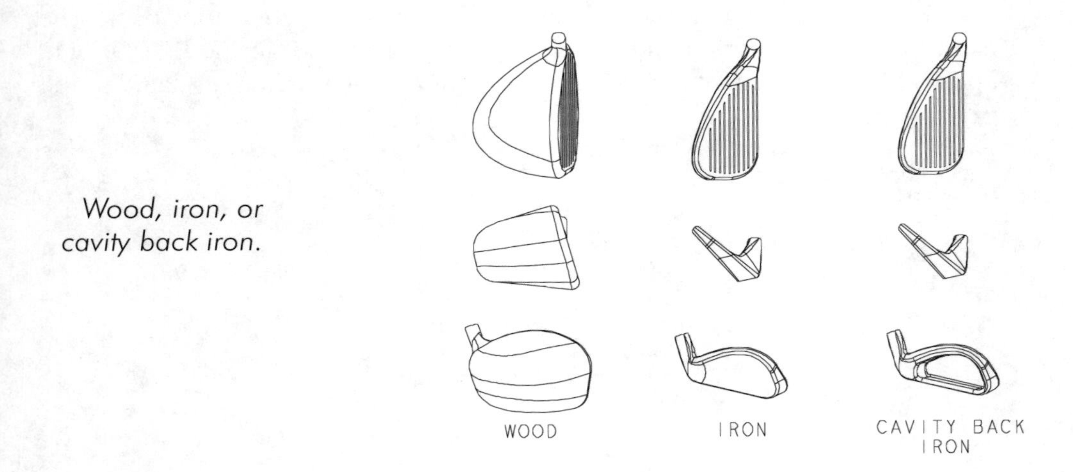

Wood, iron, or cavity back iron.

Adding Features to a Pro/PROGRAM

The entire original design intent has been captured during the development of the above Pro/PROGRAM. At this point, the sales and marketing departments decide to add another design requirement for the golf club: the appropriate club number will be applied to the sole (bottom) of the club for each club generated by Pro/PROGRAM. For example, an "8" would be applied to the sole of the eight iron.

Although it is late in the development of the programmed part, features can be easily added. The best way to add features to a Pro/PROGRAM is to use Insert Mode. Using Insert Mode will force the new feature created into the proper feature area location. However, the new inserted feature will not be fully incorporated into the program until you execute an Edit Design. The new requirement forces a unique feature applied to each club number, whether iron or wood. A conditional statement can be used for this requirement.

The breakdown for accomplishing the unique requirement is to start with the irons and work one iron at a time. The feature is added to the iron and then programmed. The following instructions begin with the three iron.

1. Regenerate the part creating a three iron.
2. Activate Insert Mode after the feature which creates the body of the club.
3. Create a cut on the sole of the golf club. Use the Text option in Sketcher to create the number three.
4. Cancel the Insert Mode.
5. Select **Program → Edit Design** and add the appropriate conditional statement for the three iron. Verify that all IF statements have a corresponding END IF statement.
6. Repeat the above steps above until all golf clubs are complete.

```
    INPUT
(ALL PREVIOUS INPUT STATEMENTS)
    END INPUT
    RELATIONS
(ALL PREVIOUS RELATIONS)
    END RELATIONS
    IF CLUB_WOOD == NO
        IF CLUB_NUMBER == 3
(Create the cut using text to build the number three.)
        END IF
        IF CLUB_NUMBER == 4
(Create the cut using text to build the number four.)
        END IF
        IF CLUB_NUMBER == N
(Create the cut using text to build the number N.)
        END IF
(Features used to create an iron are located here.)
    IF CAVITY_BACK == YES
(Features used to create the cavity back of the iron are located here.)
    END IF
    ELSE
(Features used to create a wood are located here.)
        IF CLUB_NUMBER == 1
(Create the cut using text to build the number one.)
        END IF
        IF CLUB_NUMBER == 3
(Create the cut using text to build the number three.)
        END IF
```

```
IF CLUB_NUMBER == 5
```
(Create the cut using text to build the number five.)
```
END IF
IF CLUB_NUMBER == 7
```
(Create the cut using text to build the number seven.)
```
END IF
END IF
```

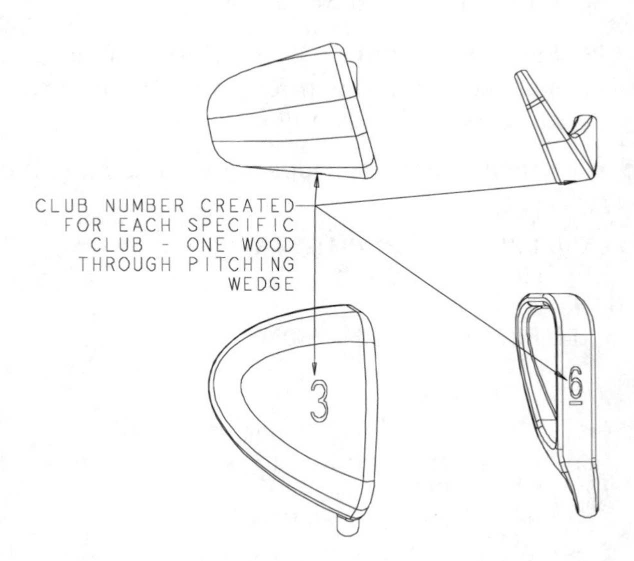

Adding the golf club's number.

Summary

Pro/PROGRAM is clearly a powerful tool for developing a variety of design problems, from a single value cube to a complete set of golf clubs. Understanding that Pro/PROGRAM allows direct access to the model's features, sequence, values, and suppression is beneficial for advanced users. In addition, understanding Pro/PROGRAM areas, and using input and conditional statements can be a tremendous asset for any Pro/ENGINEER user.

Patterning Complex Sweeps

Introduction

Patterning features or groups of features is a quick and powerful way of completing designs. However, the typical user may struggle with methods of patterning complex shapes along various geometric features. In this chapter an advanced technique for developing patterned sweep will be explored. Through an exercise, we will complete a design by incorporating the use of Datum Points on a curve, Datum Curve Split, Variable Section Sweep, Local Group, Parameters, and Relationships. The basics for the complex design problem explored below can be used for developing virtually any sweep along a complex path.

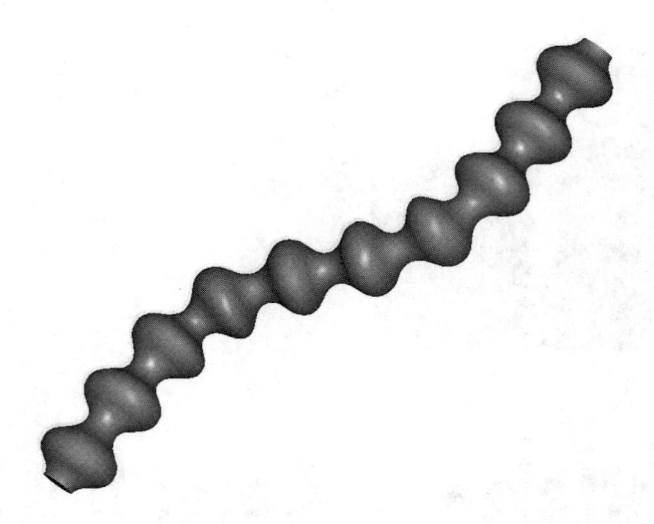

Complex patterned sweep.

Sweep Blueprint

Assume that the engineering department of XYC, Inc. has identified a need for a tube design containing several properties. Blueprint requirements appear below.

1. The length must be easily modifiable.
2. The shape or curvature must be easily modifiable.
3. The tube's center line is to be three-dimensional.
4. The internal shape is to contain bumps which must be modifiable.
5. The number of internal bumps must be patternable.

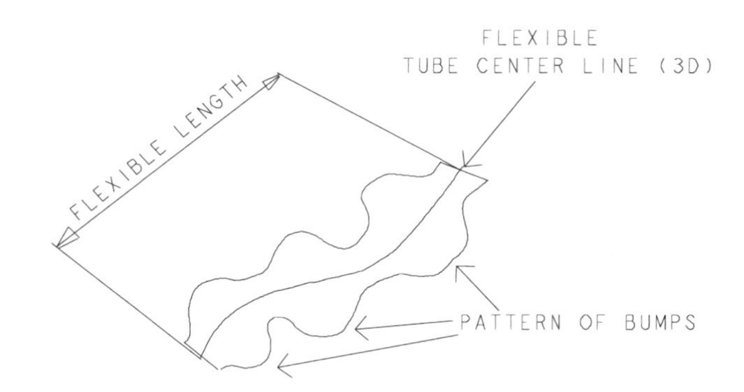

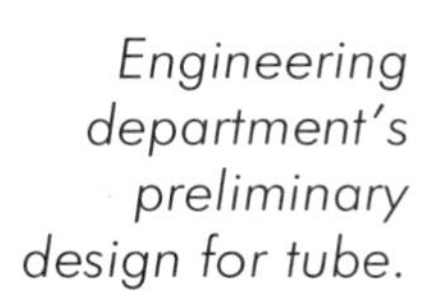
Engineering department's preliminary design for tube.

Working Through Design Requirements

As with any design the key to success is following the blueprint. The first requirement creates the starting point for designing the tube. The length of the tube, the easiest feature to control, is accomplished through a single parameter for the total length. The two-dimensional shape for the tube can easily be made flexible through the use of a spline. Create a Datum Curve by sketching it on a plane.

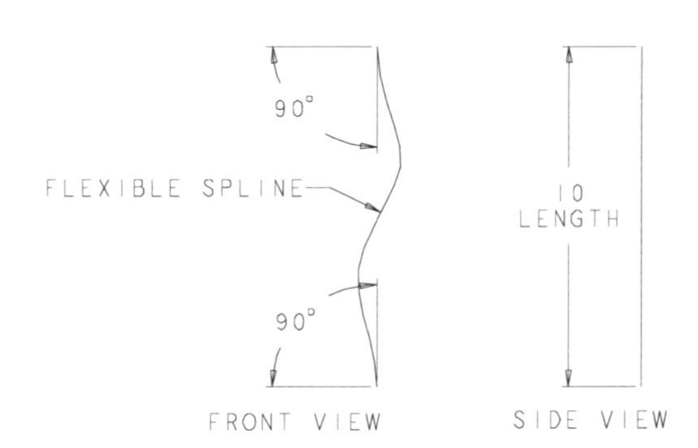

The first design requirement.

Note in the above illustration that the ends of the spline have been created with tangency. The tangency value is equal to 90 degrees, which adds control for the curve. The tangency condition will allow for the start and end of the tube to be modified to different angles later in the design, if necessary.

3D Center Line

Creating a three-dimensional center line for the tube is the next step in design development. To create a three-dimensional center line two additional features will be created. The first is a surface, and the second, a projected curve. The reason for the surface and projected curve is to permit easy identification for future changes. Simply put, the curves and surface will become the base or foundation of the tube.

The surface will be created using a sketch similar to the one used to create the first curve. Using a spline once again allows for complete flexibility of the shape. In order to develop the proper three-dimensional curve to be projected in the next step, the extruded surface will be created perpendicular to the first curve. The end points will be aligned to the end points of the first curve. The alignment ensures that a single dimension controls the total height of the tube.

Extruded surface perpendicular to first curve.

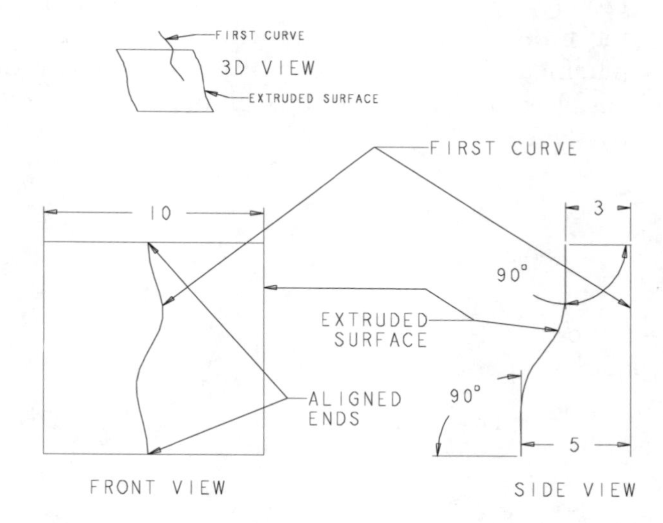

The actual curve to be used as the center line of the tube is created by projecting the first curve onto the extruded surface. The result will be a three-dimensional curve that can be controlled by changing the first curve or the extruded surface.

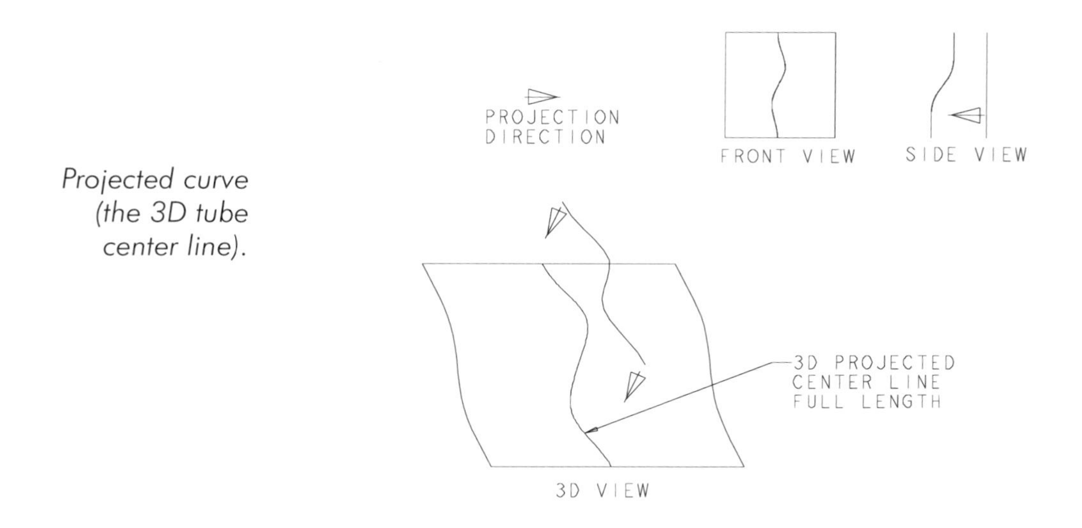

Projected curve (the 3D tube center line).

Splitting the 3D Center Line into Portions

Splitting the three-dimensional center line into portions represents an interesting problem. At this juncture, a review of design elements may be appropriate.

- ❑ The bumps will later be created or defined to follow a portion of the three-dimensional center line.
- ❑ Each portion must then be patterned to fit the total length of the three-dimensional center line.

The key to this part of the design problem is the breakdown of the three-dimensional center line. The center line must be split into a portion that can easily be swept along and then patterned.

To accomplish the splitting of the three-dimensional center line, datum points will be necessary. The datum points are created using the following option: **On Curve → Length Ratio**. Two points are necessary and can be created as a single feature as follows:

Feature → Create → Datum → Point → On Curve → Length Ratio

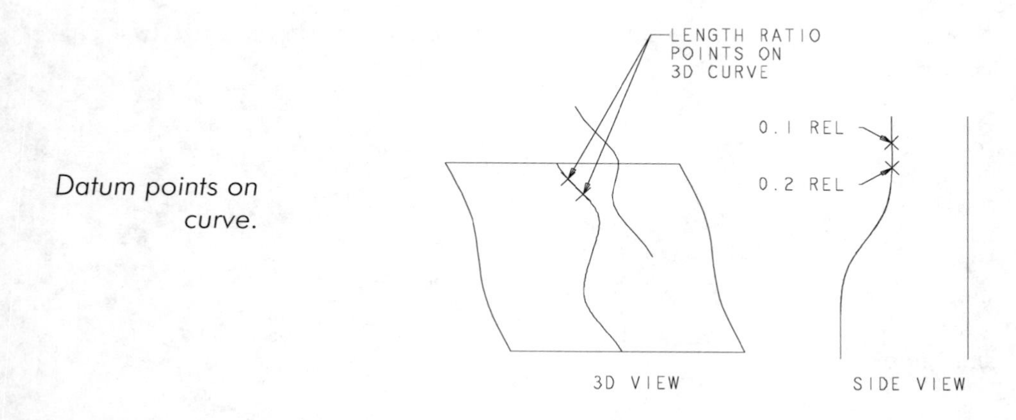

Datum points on curve.

✔ **NOTE:** *Pro/ENGINEER offers many methods of creating datum points on a curve. The other techniques (Offset and Actual Length) may also prove to be valuable during development of a similar product.*

The two points have been created with a ratio between them of one-tenth the total length of the three-dimensional curve. With that information established, the curve will now be split into one of the portions to be patterned later. Using **Create → Datum → Curve → Split**, select the three-dimensional center line and then the first point. Pro/ENGINEER will prompt you for the side of the curve you wish to keep. Repeat this procedure using the second point while maintaining the portion of the center line between the two points.

Creating the bump's center line portion using Split.

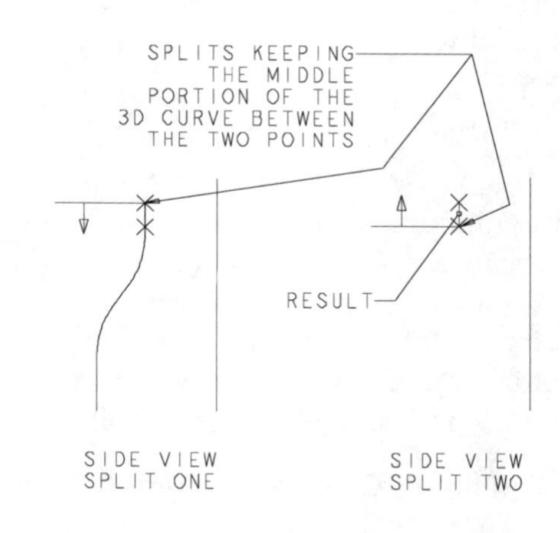

Once the three-dimensional center line has been split into a portion, flexing the values will test the movement of the portion. This test will help to ensure successful patterning later in product development. Modifying the values of the first and second point from 0.1 to 0 and 0.2 to 0.1 will move the split portion to the start of the original three-dimensional center line. Modifying the values from 0 to 0.9 and 0.1 to 1 will move the portion to the end of the original three-dimensional center line. Complete testing by moving the points back to 0 and 0.1, respectively.

Testing (flexing) the curve portion for patterning.

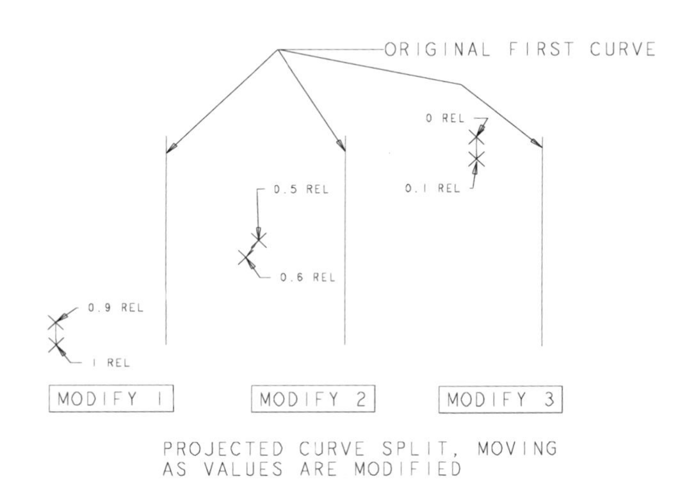

Creating the Bump to be Patterned

The creation of the bump will be accomplished using a Variable Section Sweep. This type of sweep allows unique control of the cross section used in the sweep. The basic cross section for the tube is a simple circle dimensioned with a diameter. However, the value of the diameter will vary along the center line (spine) of the sweep. The variance of the diameter from small to large and back to small which creates the bump, will be accomplished by driving the value via a Graph feature. Modifying the shape or values of the Graph will then update the sweep, lending the design tremendous control.

The Graph feature to be used in the Variable Section Sweep must be created first. Creating the Graph feature permits the sweep to read or be driven by the graph. Another option is to create the sweep, and then use Insert mode to place the graph before the sweep. Select **Feature → Create → Datum → Graph** and enter a name for the graph. After entering a name (*BUMPS*), Pro/ENGINEER will activate the Sketcher. Sketch a cross section which defines values for the diameter of the sweep.

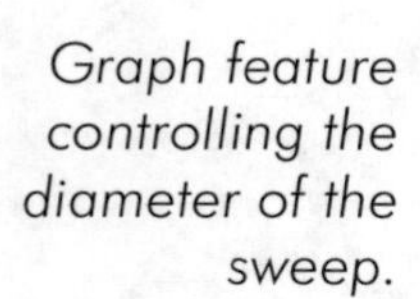

Graph feature controlling the diameter of the sweep.

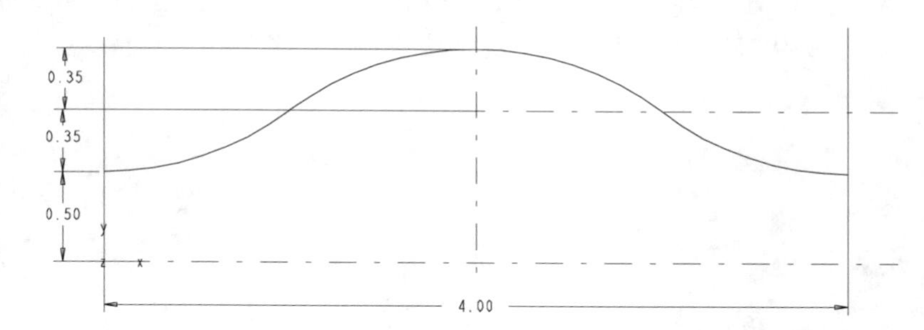

✔ **NOTE:** *The graph feature must contain a sketcher coordinate system. The values the graph outputs for the Variable Section Sweep will be based off the X values for the diameter and Y values for the location along the sweep. In addition, when the Graph is complete, make a note of the symbolic value for the length of the graph (D459) for use in the Variable Section Sweep relation.*

Next, create the Variable Section Sweep by selecting the following commands:

Feature → Create → Protrusion → Advanced → Var Sect Swp → Nrm To Spine

Select the center line portion as the "spine" and select the first curve as the "x-vector" curve. The two curves will complete the selection for the Variable Section Sweep. Pro/ENGINEER will then automatically switch into the Sketcher for defining the cross section. The cross section will be a simple circle with its center lying on the end of the spine (center line portion) curve. Before the sweep is completed, a relationship must be created to link the value of the diameter to the values from the graph.

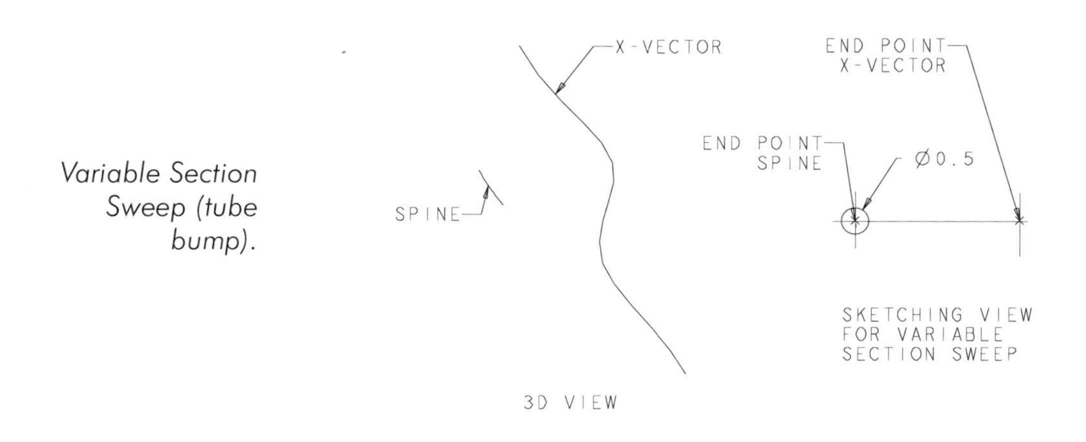

Variable Section Sweep (tube bump).

To create the relationship to the Graph feature, modify the value for the diameter while in the Sketcher and enter the following:

```
EVALGRAPH ("BUMPS", TRAJPAR * D459)
```

- ❑ BUMPS is the name of the Graph feature.
- ❑ TRAJPAR is a value from zero to one along the Graph.
- ❑ D459 is the dimension symbolic value for the length of the Graph.

Regenerate the section and select Done. Once the sweep is complete, the geometry will reflect the Graph feature.

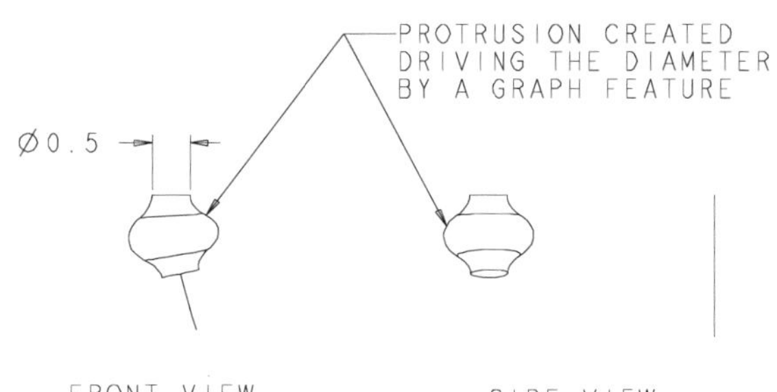

The completed bump.

Relating Design Variables for a Successful Pattern

The final steps involve creating the pattern, setting up parameters, and relating the design variables. Before relating the variables, a bump pattern will be created by using the Local Group option. This option allows for the patterning of more than one feature. Select **Feature → Group → Local Group**, and enter a name for the group (*BUMP_PATTERN*). Next, the features to be contained within the group should be selected. The Local Group consists of the following:

- ❑ Points used to split the three-dimensional center line.
- ❑ Curve Splits (two features).
- ❑ Graph feature (BUMPS).
- ❑ Variable Section Sweep.

Once the features for the group have been selected and the group created, the pattern is created. While still in the Group menu, select Pattern, and then pick the group that was just created. Picking anywhere near the group will select the group since Pro/ENGINEER seeks only groups during this pick. Pattern the Local Group using the length ratio point values for the pattern. Pro/ENGINEER will then prompt you for pattern increment. Enter the same length ratio value from the original points for the pattern increment (0.1 for both points). This value is important because the graph length value is currently tied only to the current ratio.

✔ **NOTE:** *The pattern feature will prompt for values to pattern in the first direction, and then the second direction. It is important to select both length ratio values for the pattern in the first direction and none in the second. This selection will ensure that the group patterns correctly.*

Pro/ENGINEER will then prompt you for the number of patterns you wish to create. Entering 10 patterns will complete the tube based on the current length ratio of 0.1.

Patterning the Local Group by length ratio points.

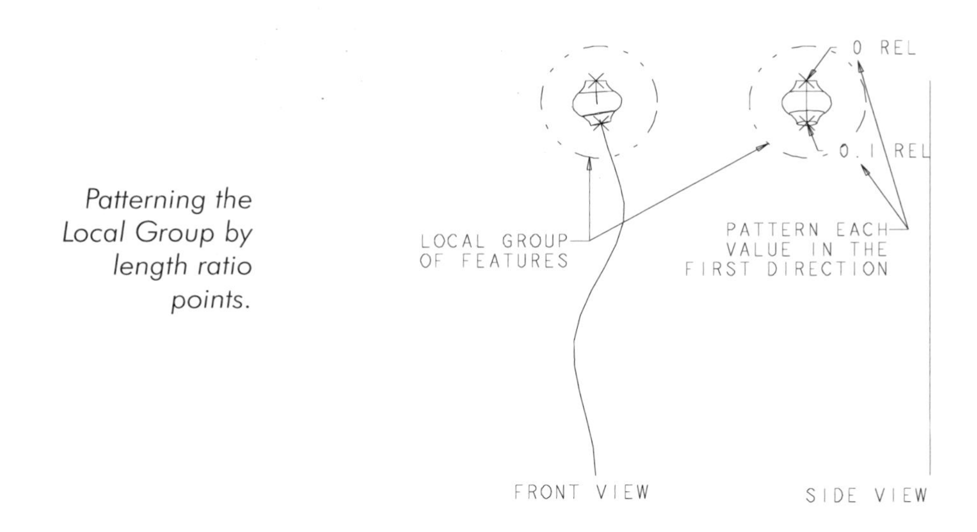

Local Group patterned ten times.

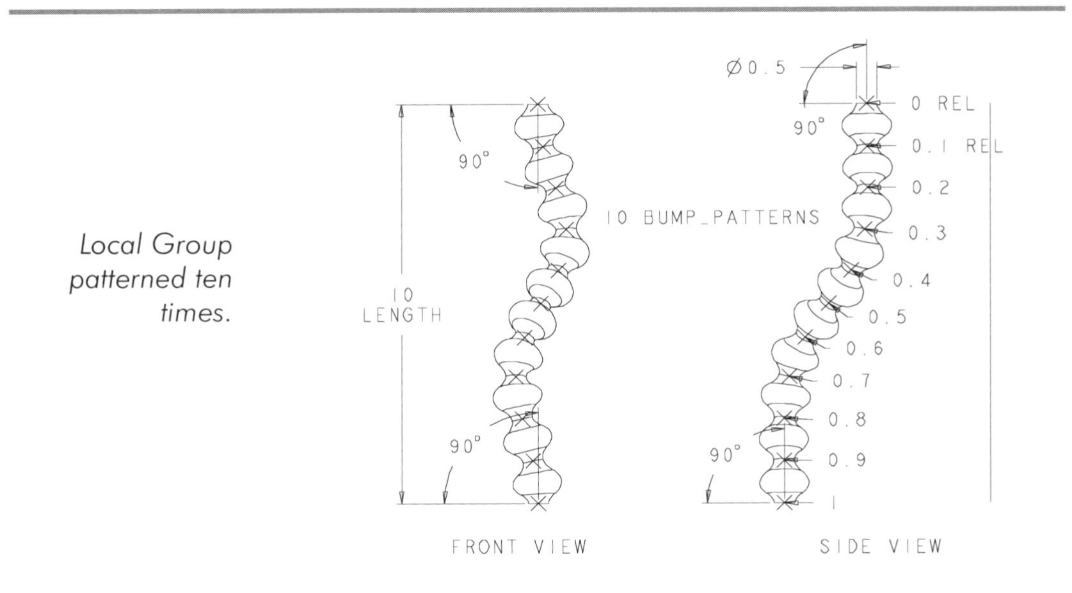

! **WARNING:** *The current design will look great, but without establishing additional relationships or time-consuming modifications the design will not meet requirements if any changes occur. To resolve this issue and accommodate the design intent for flexibility, the design variables will be related.*

To wrap up the design and make a powerfully patterned part, all necessary design variables will be related. The first design variable involves the true length of the projected curve. This curve establishes the total curve length for the points created on the curve using Length Ratio. Our design intent is for the number of patterned bumps to determine the actual length ratio. Two parameters will be created which will represent the value for the ratio of the pattern and number of bumps desired on the tube. Select **Set Up → Parameter → Create → Number**. Respond to Pro/ENGINEER's prompts as follows:

```
Enter the name of the parameter:
      NUMBER_BUMPS
Enter the value for NUMBER_BUMPS:
      10
```

Repeat for the bump ratio value.

```
Enter the name of the parameter:
      BUMP_RATIO
Enter the value for BUMP_RATIO
      0.1
```

After the parameters are created, the necessary relationships to complete the part can be created. The relationships required for total part flexibility consist of the following:

- NUMBER_BUMPS = P9 (P9 is the symbolic value for the Local Group pattern.)
- BUMP_RATIO = 1 / NUMBER_BUMPS (This value will establish the proper value to increment each curve split portion during patterning.)
- D418, D494, and D495 = BUMP_RATIO (This relationship establishes the proper values for each length ratio point which splits the curve during patterning.)

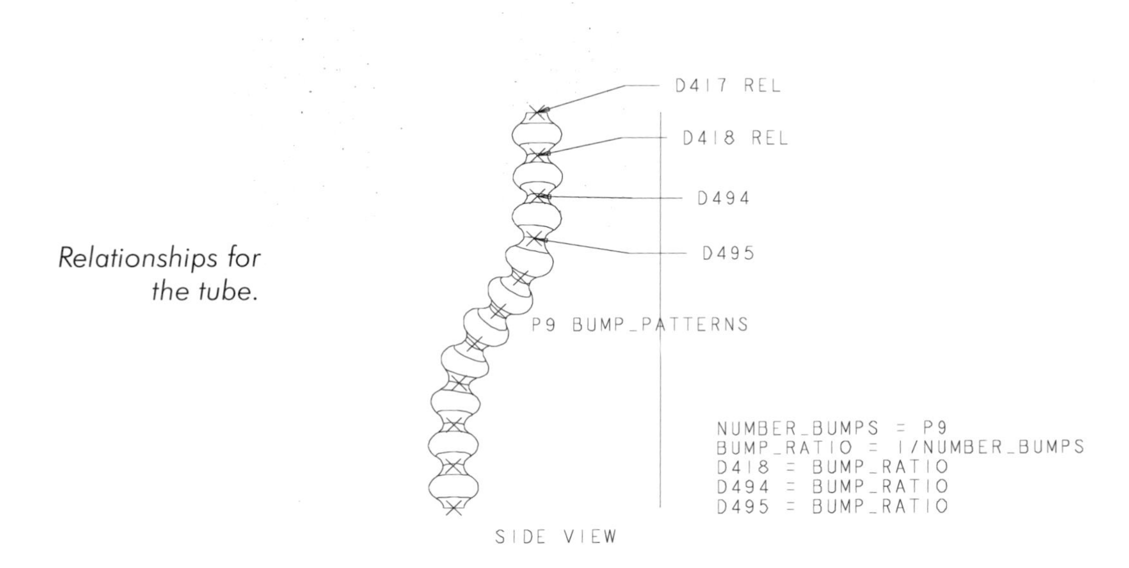

Relationships for the tube.

The part is complete and ready for many different sizes and shapes based on the original design requirements. Flexibility is built into the model.

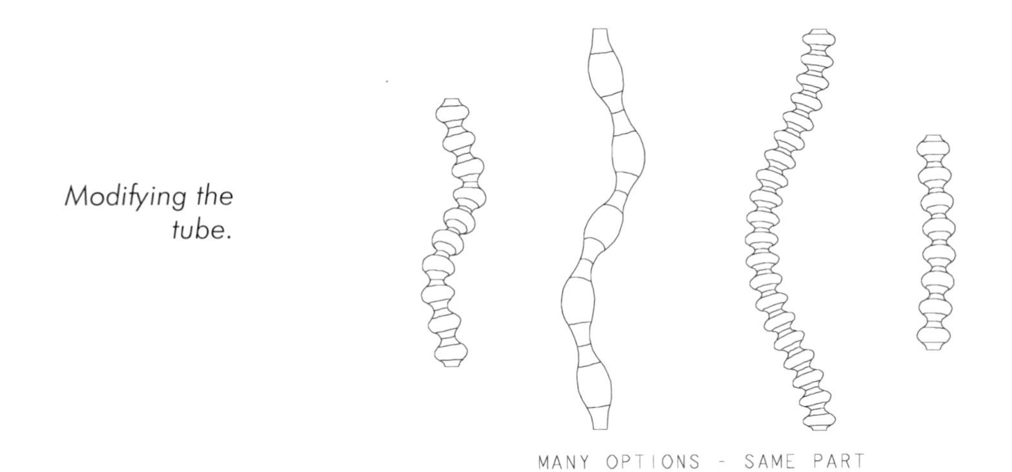

Modifying the tube.

Summary

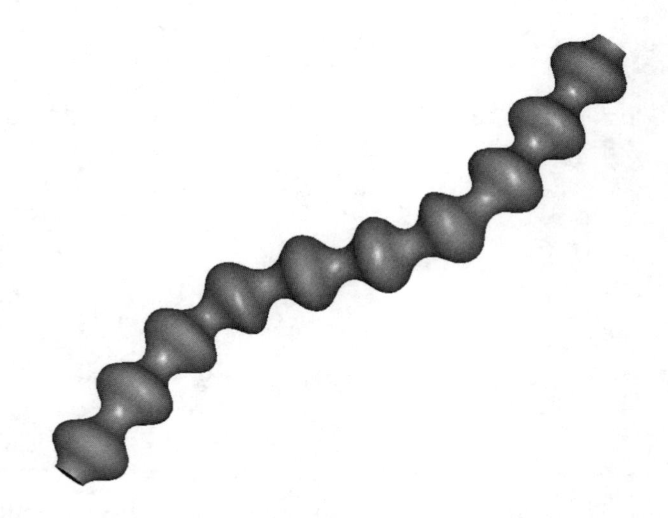

Complex pattern sweep (the completed tube).

The tube with multiple bumps demonstrates how a complex pattern can be developed. Using advanced techniques such as the Length Ratio point option in conjunction with the Variable Section Sweep, Local Group, and relationships drove the development of the complex pattern sweep. In addition, the design blueprint, a few tests of the desired outcome as the part is being built, and good techniques help solve the complex tube pattern. These techniques show the amount of flexibility that can be built into a design as well as establishing a base for a complex pattern sweep.

Creating a Bending Spring

Introduction

Pro/ENGINEER offers a very powerful method of creating springs through the use of the Helical Sweep feature. Although this method is powerful, its application is restricted to straight sweeps along an axis. In this chapter, we explore an advanced method for creating a spring. This technique will illustrate the use of Variable Section Sweep and Graphs to create a spring which bends. First, the blueprint and the goals for the design will be established. The problem will then be explained in a step by step fashion resulting in a complex bending spring.

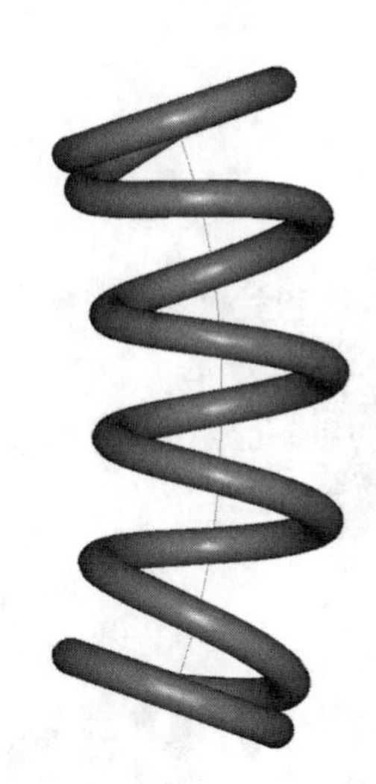

The bending spring.

Blueprint for the Bending Spring

Assume that our bending spring problem centers around the need for the design and engineering departments to create a spring representative of its function. The function of the spring is to bend as it is compressed through angular travel. The bending spring has many requirements which must be fulfilled before its function is achieved. To describe the problem, assume that the customer and the engineering department have developed the example sketch appearing in the next illustration.

Engineering example of bending spring problem.

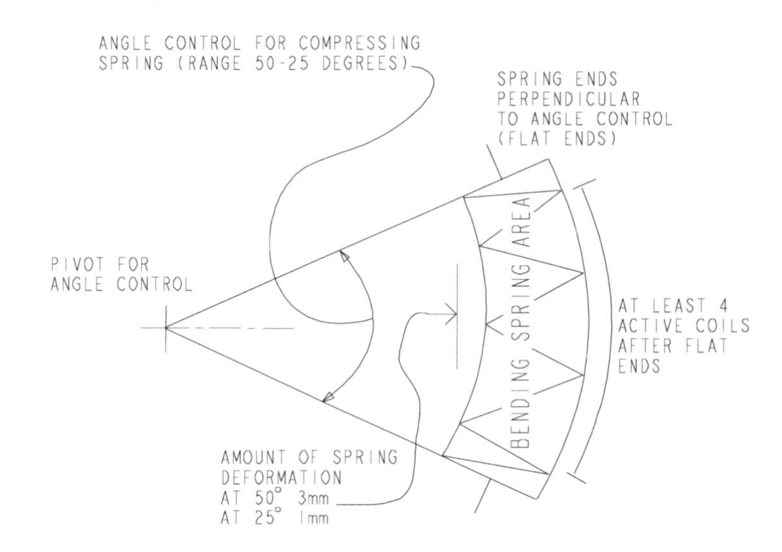

With the use of known information on the strength of a spring and the customer's objectives, the engineering department has laid out four items which establish the blueprint for the advanced spring design.

1. The spring must be able to move from a starting angle of 50 degrees at a free length to an angle of 25 degrees compressed.
2. Based on the known strength of the spring, the amount of deformation (bending) will be 3mm at 50 degrees to 1mm at 25 degrees.
3. The spring is to have flat ends to accommodate subsequent assembly requirements.
4. There must be at least four active (compressing) coils between the flats of the spring.

Setting up the Bending Spring

Once the blueprint is established, the next step is to set up the first feature that will help to achieve our objectives. The features for the bending spring will initially be quite simple. However, the simple features will become the foundation for the complex bending spring.

To create the first feature for the bending spring, we will focus on the first requirement of the blueprint. The first step is to establish a desired movement for the spring from 50 degrees to 25 degrees. This desired movement is accomplished by creating a Datum Curve feature. The datum curve is sketched on a set of default datum planes.

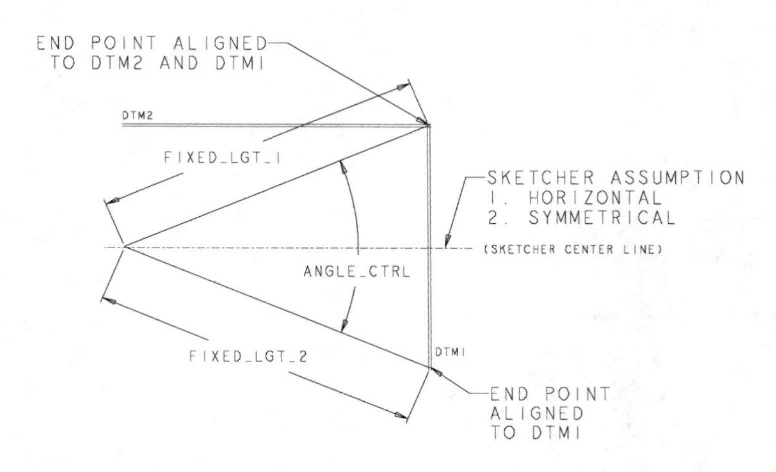

Angle control for the bending spring using Datum Curve.

✔ **NOTE:** *The creation of the datum curve takes advantage of Pro/ENGINEER's Sketcher assumptions and alignments to help minimize dimensions.*

Two additional datum curve features will be created to complete the foundation for the complex bending spring. Both datum curves are created in such a manner that they relate directly to the first datum curve created (the angle control curve).

The first of the two additional curves will be sketched as a three-point spline connected to both ends of the angle control curve with tangency. The spline will later serve as the "spine" for the Variable Section Sweep. In addition, the curve will help fulfill the second blueprint requirement. The curve will have a dimension to be discussed later for controlling how much deformation (bending) the spring has as the angle control changes.

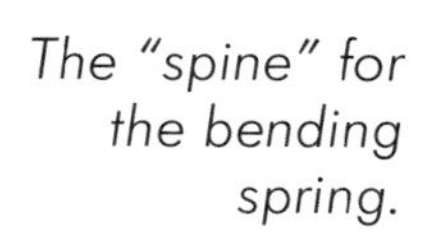

The "spine" for the bending spring.

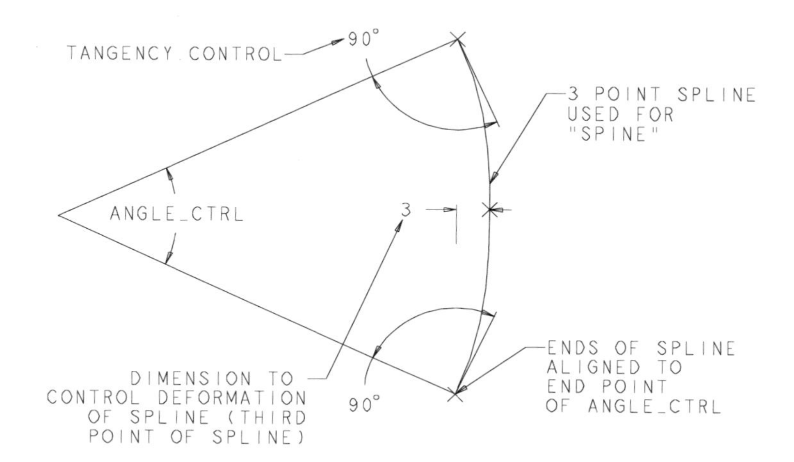

The second additional curve will be sketched as a straight line. This line is created to complete the curve foundation for the complex bending spring. As a straight line this curve will become the "x-vector" for the Variable Section Sweep.

The "x-vector" for the bending spring.

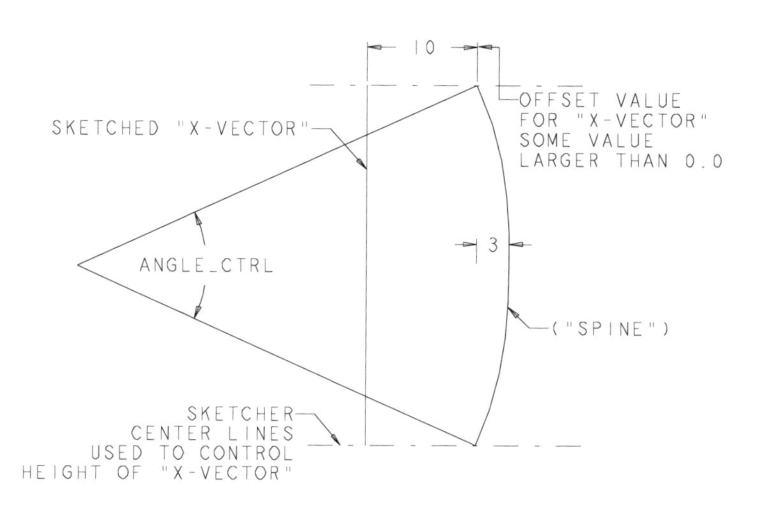

Developing the Foundation for the Bending Spring Shape

After completing the curve foundation using datum curves the focus of the development shifts to shape of the spring. As with any spring the basic shape is helical. However, the complex bending spring incorporates flat ends, bending, and a specific number of coils into its shape. Like the datum curves these features expand the foundation on which the actual spring will be created.

Adding in specific shape control for the spring is easy if the spring is developed using a Variable Section Sweep. Additional control is accomplished by using Graph features and relationships. Each Graph feature then drives the sweep to the desired shape.

The next feature created will be a Datum Graph which will create the flat ends and the number of coils for the spring. The following command sequence will initiate this development: **Feature → Create → Datum → Graph**. Pro/ENGINEER will prompt you to enter the name of the graph (*HELIX_GRAPH*). Next, the Sketcher will be activated, allowing for the creation of the graph.

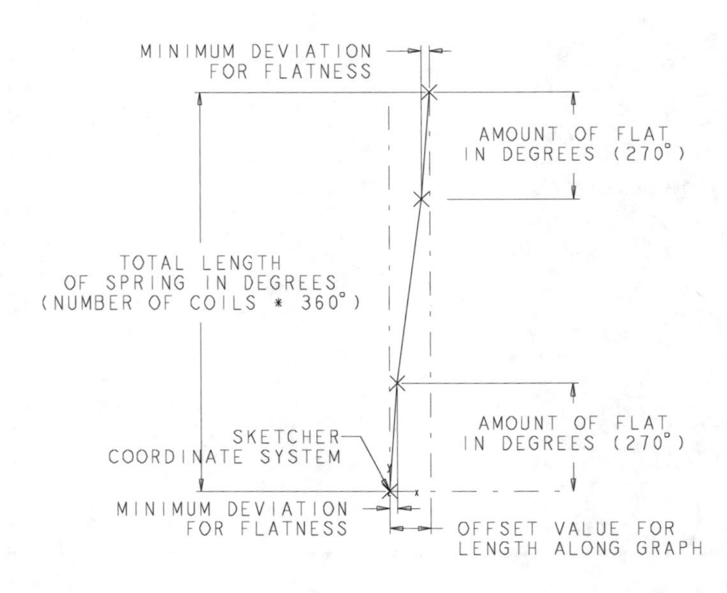

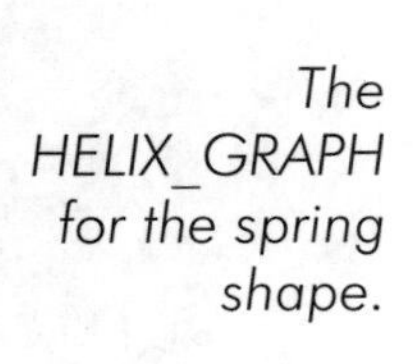
The HELIX_GRAPH for the spring shape.

✔ **NOTE:** *Upon using Variable Section Sweep, there is no method to make the spring exactly flat. Note that in the HELIX_GRAPH the spring actually drops 0.1mm as the coil rotates over 270 degrees. In general, this amount of deviation from perfectly flat is acceptable.*

An additional graph will be created to drive the amount of deformation of the spring at any of its angle positions (50 degrees deform 3mm and 25 degrees deform 1mm). This Graph feature will be created in the same manner as the previous feature. Use the name *DEFORMATION_GRAPH*. The graph will establish that at 50 degrees the deformation will be some variable (3mm) and at 25 degrees another variable (1mm). Any variable between the 50 and 25 degrees will also automatically drive the deformation to its appropriate value. For example, at 47.5 degrees the deformation will be 2mm.

The DEFORMATION_GRAPH for the spring.

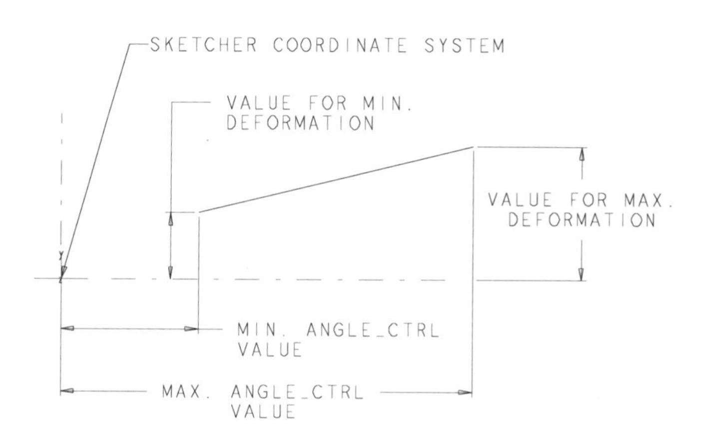

Using Setup and Foundation to Generate the Bending Spring

After the setup and foundation have been created, generation of the actual bending spring can begin. The creation of the spring will use the previous groundwork (datum curves and graph features) and tie in relationships to generate the complex bending spring.

The initial feature to develop the spring will be a surface created using Variable Section Sweep. The resultant edge of the surface will then be used to create the solid spring. To create the surface select the following commands:

Feature → Create → Surface → New → Advanced → Done → Var Sec Swp → Done → Norm to Spine → Done

The next options refer to the selection or creation of trajectories for the sweep.

✘ ***TIP:*** *The trajectories may be created on the fly at this point. There are many items, however, for control of the bending spring. Many advanced users will often create the trajectories ahead of time using Datum Curves such as in the current example. Creating the trajectories as curves allows for easy access and control. Use your experience and complexity of the problem at hand to judge whether or not to create the trajectories on the fly.*

To create the bending spring, select the trajectories. The first curve selected for the Variable Section Sweep will be the "spine." The second selection for the Variable Section Sweep will be the "x-vector." These two selections will complete the required trajectories for the creation of the surface.

Pro/ENGINEER will then place you in the Sketcher looking down the end of the "spine" curve. The goal is to create a surface which will "spine" as it is swept along the "spine" trajectory. The spinning of the surface is generated by creating a Sketcher relationship between an angle in the sketch, the length the sketch moves along the curve (*trajpar*, or trajectory parameter), and the *HELIX_GRAPH* graph feature.

When the Sketcher relationship is complete, regenerate the sketch. The sketched curve will snap to its starting angle for its sweep (0 degrees). Select Done and review the completed surface.

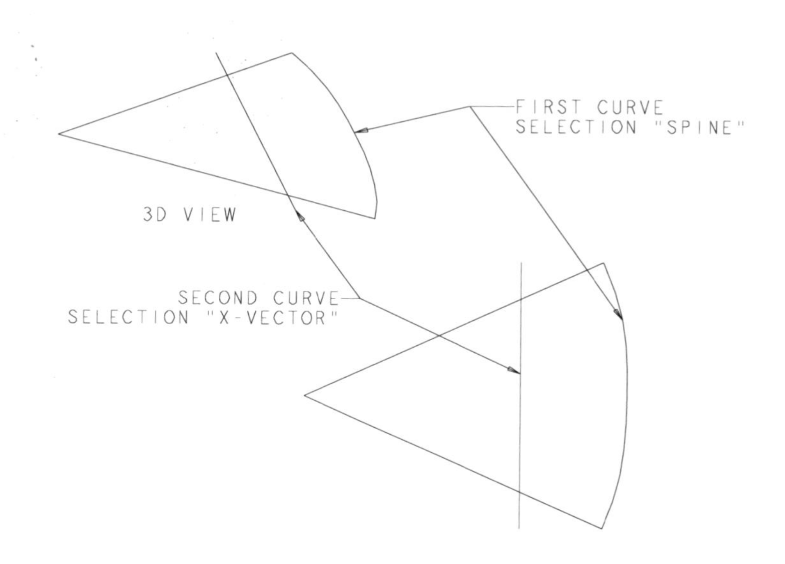

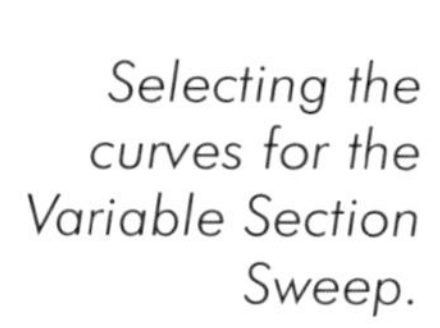

Selecting the curves for the Variable Section Sweep.

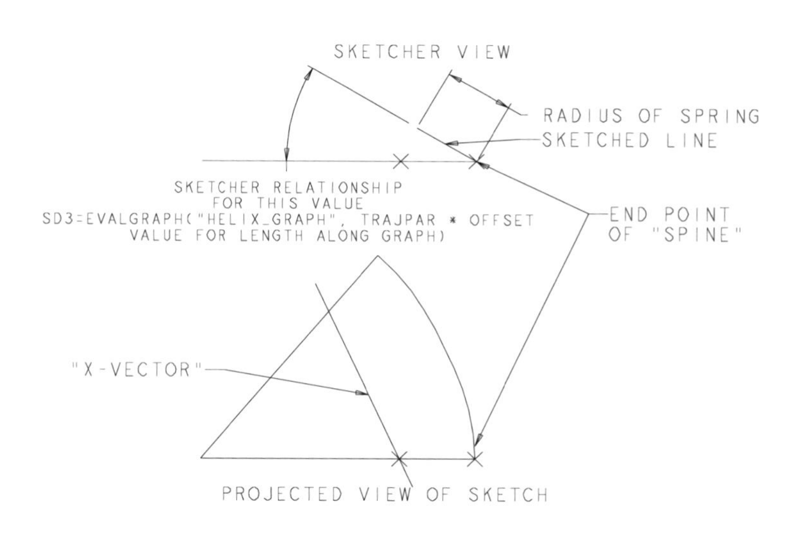

The sketch and relationship for the bending spring's surface.

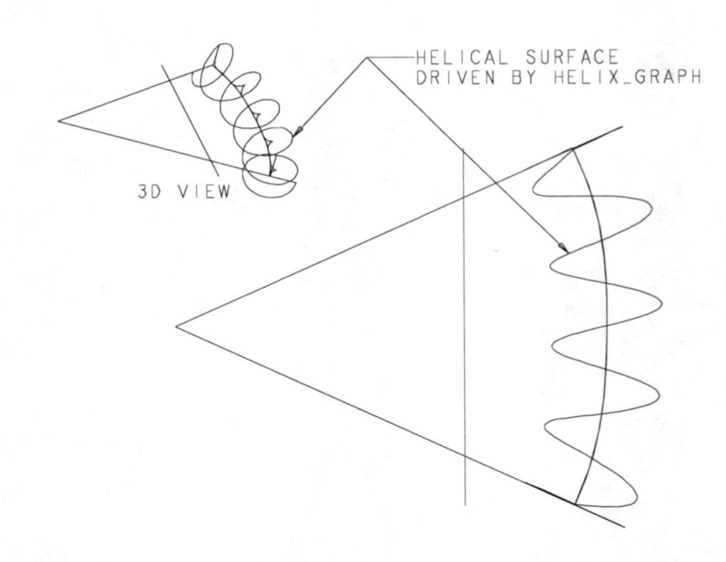

The completed bending spring's surface.

✘ **TIP:** *When creating a spring or a helical shape using Variable Section Sweep, right-hand or left-hand rotation (spinning) may be important. If the Variable Section Sweep is to be a right-hand rotation, the sketch should be completed so that the angle for the rotation is above the spine and x-vector, and vice versa for a left-hand rotation.*

SKETCHER VIEW

RADIUS OF SPRING
SKETCHED LINE

RH SPIN PARAMETER

VARIABLE SECTION SWEEP IS
CREATED SWEEPING TOWARDS YOU.
THE ANGLE VALUE IS THEN INCREASED
BASED ON THE VALUE OF THE TRAJPAR
AND THE GRAPH.

LH SPIN PARAMETER

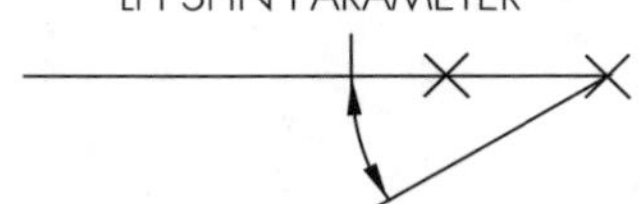

Right- and left-hand Variable Section Sweeps.

A final relationship is created to force the deformation to change as the angle control parameter is changed. Incorporating this control is done through a relationship of the deformation dimension to the deformation graph. The deformation dimension will be driven directly from the parameter (value) extracted from the deformation graph. The value read from the graph will be dependent on the current angle control. For example, at 50 degrees the value for the deformation dimension will be 3mm.

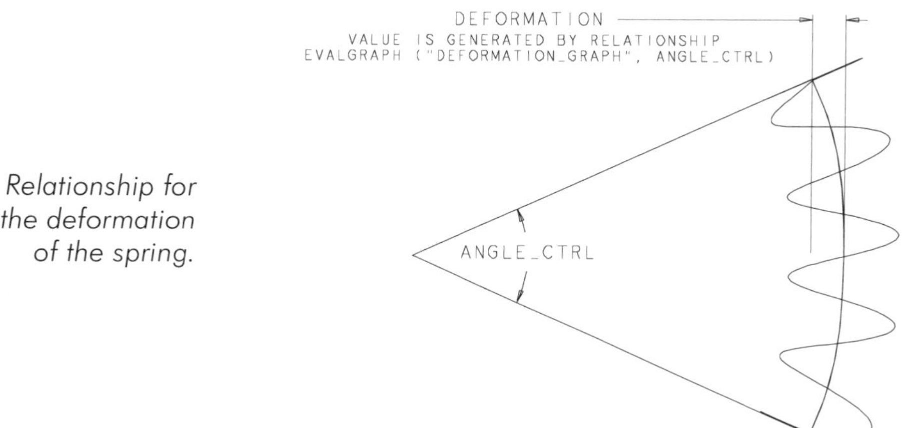

Relationship for the deformation of the spring.

Creating the Solid Portion of the Bending Spring

To complete the design, a **Protrusion → Sweep** will be added. The Sweep will use the edge of the surface from the outside top to the outside bottom as its trajectory. After selecting the edge of the surface for its trajectory, sketch a circular cross section for the spring's shape.

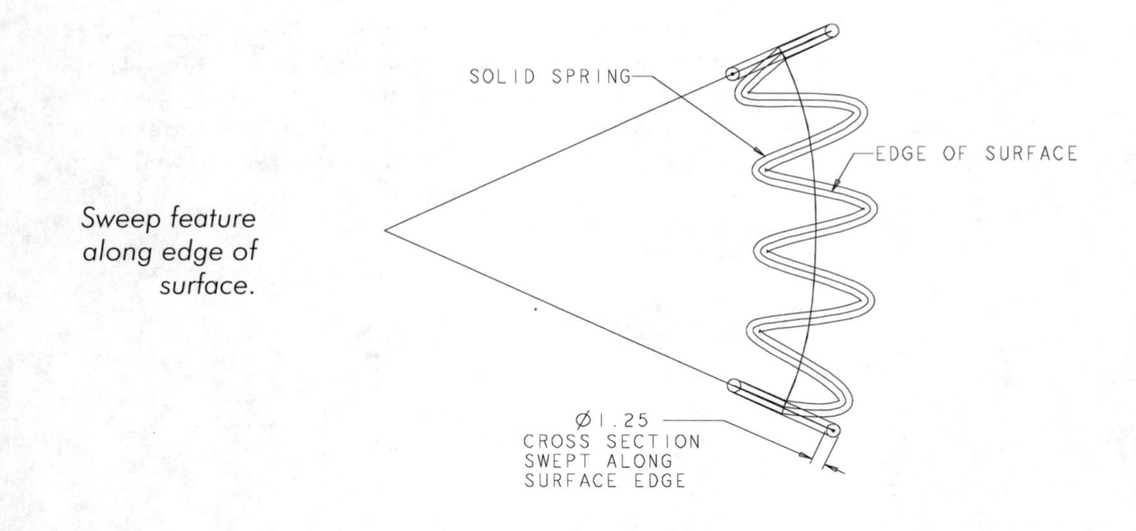

Sweep feature along edge of surface.

After completing the Sweep, Layer and Blank the x-vector and the surface to clean up the design.

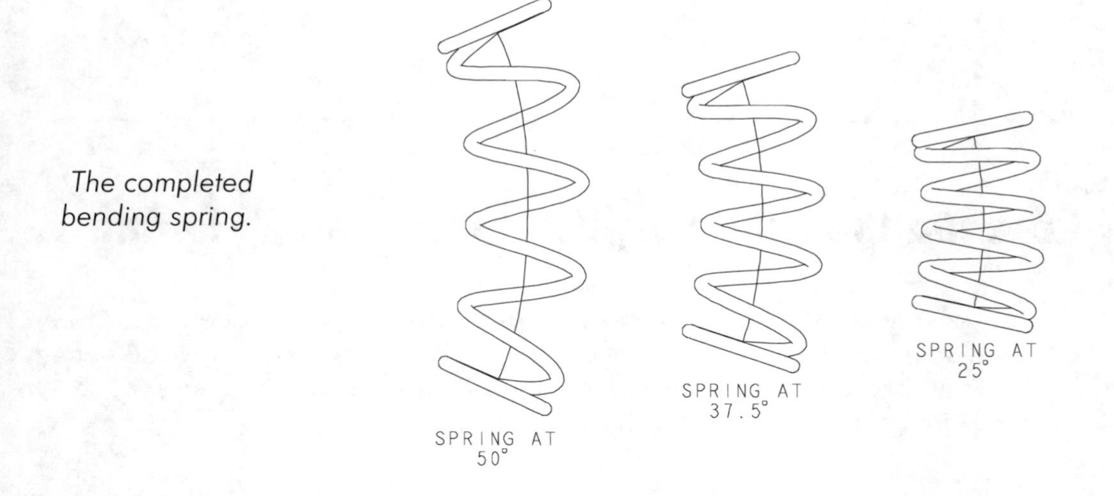

The completed bending spring.

Summary

In this chapter an advanced method for creating a spring was explored. The technique allows you to go beyond the traditional methods and limitations of creating a spring. This technique illustrated the use of Variable Section Sweep in concert with Graph features to create a spring which bends. The blueprint and the goals for the design were established. The problem was then systematically solved by breaking down the design into basic features to build a sound foundation aimed at yielding a complex bending spring.

PART Three

Free-Form Design

Advanced Curves and Surfaces

Introduction

This chapter provides an introduction to advanced complex curve and surface design within the Pro/ENGINEER environment. Pro/SURFACE functionality is incorporated to demonstrate how parametric surface modeling interacts with solid modeling. Tips and techniques are utilized when focusing on body design principles and practices that include master sections, patch structure, and slab theory.

We also cover basic mathematical definitions of curves and surfaces, how they are applied in the design blueprint, and how they are evaluated for smoothness, tangency, and curvature continuity. These definitions will help you better understand how curves and surfaces are related in the formation of parametric and free form design.

Curves

Pro/ENGINEER supports many types of curves such as (1) straight lines; (2) conic sections (circles, ellipses, parabolas, and hyperbolas); (3) free-form curves (B-splines and nurbs); and (4) surface intersection curves. Although the detailed mathematical definitions of these curves are beyond the scope of this book, it is important to understand the basic concepts. Conics and splines are covered because they are important in free-form design. Development of these entities is primarily driven by the aerospace and automotive industries.

Conic Sections

Conic sections were first described in Plato's time as curves in which a plane intersects a cone. Today these curves are described as second order quadratic equations located in a coordinate plane. Conic curve sections produce perfectly smooth curvature accelerations and are typically used in advanced shape design. These curves primarily focus on defining large slab areas. The four conic section types are shown in the following illustrations.

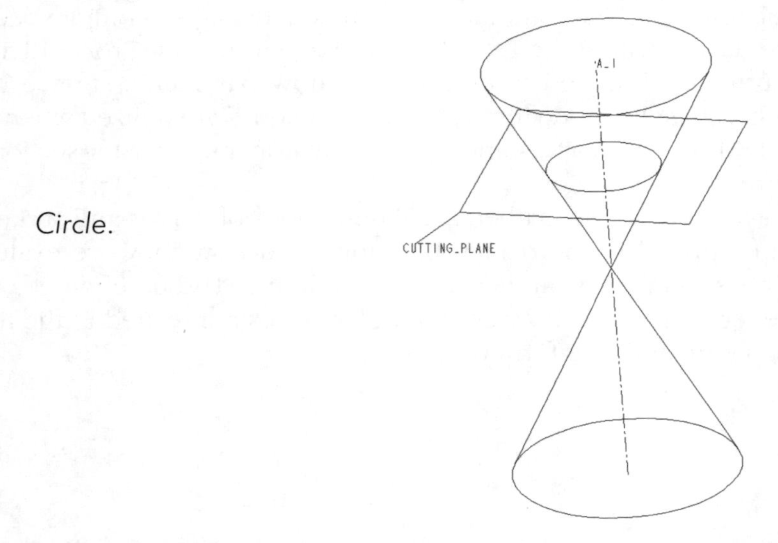

Circle.

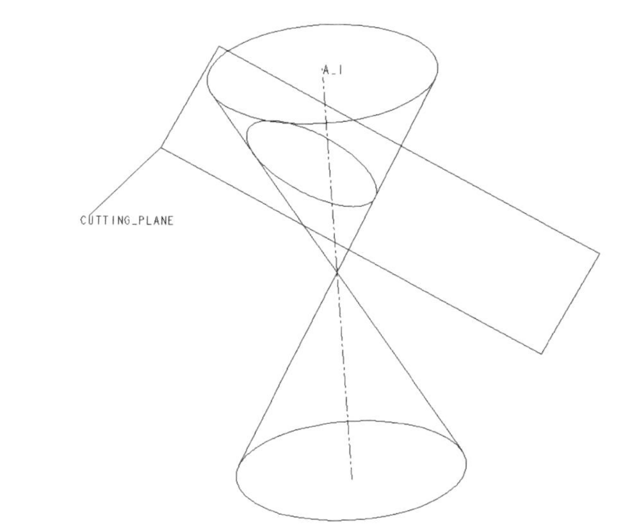

Ellipse.

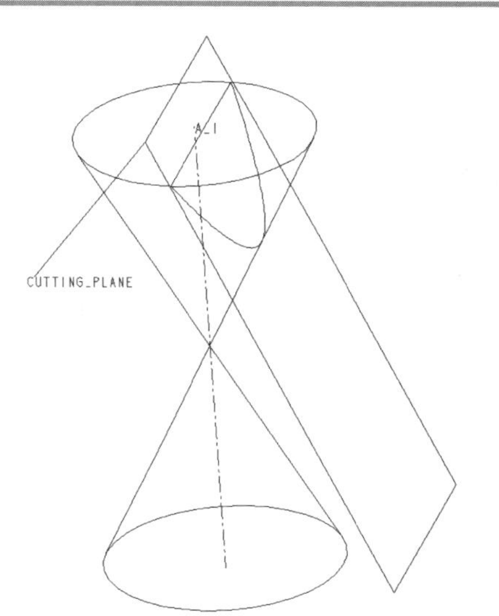

Parabola.

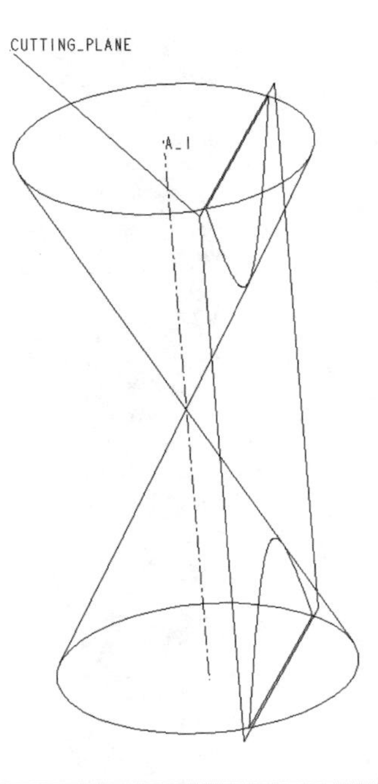

Hyperbola.

Creating Conics

Conics are created in Sketcher mode under the Adv Geometry menu. Two points with a Rho value (Shape Factor) or three points are needed to define the curve. The slope (Angular Dimensions) of the conic must also be defined or aligned to a datum axis. An example layout follows:

```
Rho is the ratio of BC/AC
Parabola = .5 (Default)
Ellipse = .414 (Square root 2 over 2)
Hyperbola = .75
```

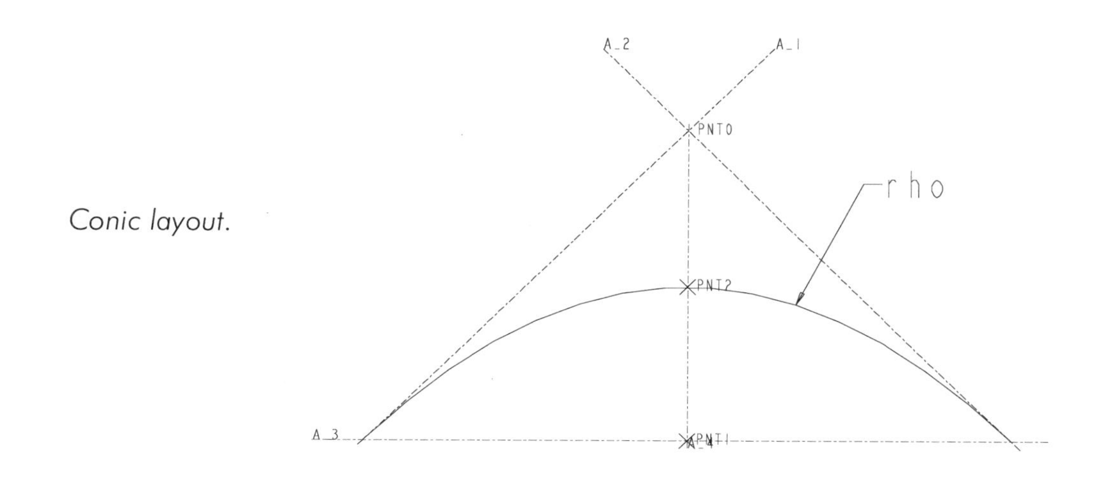

Conic layout.

Splines

Splines are curves that run smoothly through an empirical set of points. Various techniques and methods, including "parametric curve definitions," have been employed over the last 20 years to mathematically define these smooth shapes. Pro/ENGINEER employs all techniques for defining a spline curve (cubic spline, nurbs, and so forth). In mathematical terms, a B-spline type curve is described by the following parameters:

- ❑ Degree of the polynomial arc forming the curve. Degrees range from 1 to 15 with the order being equal to (n + 1).
- ❑ Knot sequence. Uniform and non-uniform sequencing at the junction of the arcs. Knot sequence is concerned with continuity blending.
- ❑ Control Poly. Lies off the defining arc and is used to influence the shape of the curve. Each arc of the curve is contained in the convex envelope of the control points defining the arc.

Creating Splines

Splines consist of two end points with or without intermediate pionts, and are are created in Sketcher mode under the Adv Geometry menu. You create a spline by selecting the spline point path or the control polygon that controls the spline. Both ends must be fixed (dimensioned or aligned) with or without tangency and curvature constraints. Tangency and curvature constraints may be added later by modifying the spline.

✔ **NOTE:** *You are initially given the option to create the spline with no tangency constraints or at the beginning, at the end, or both.*

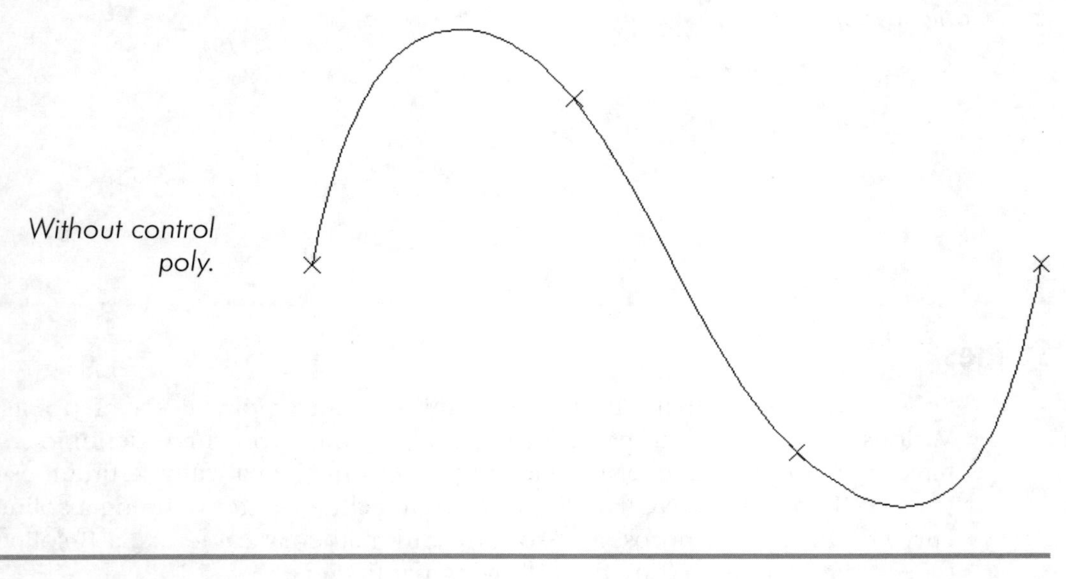

Without control poly.

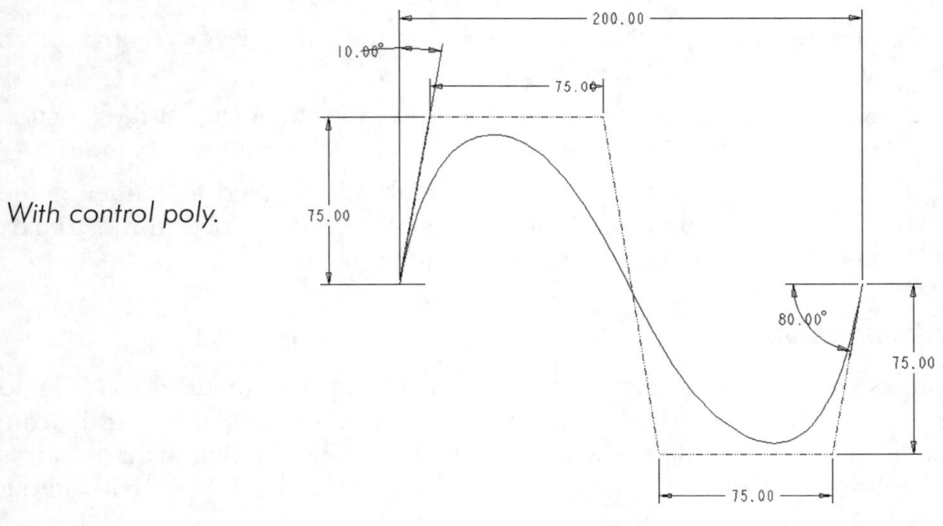

With control poly.

Constraints and Options

To create splines, select the Adv Geometry menu in the Sketcher. Keep the following in mind when creating splines:

- Indicate the desired spline path points.
- End points may or may not have tangency constraints.
- Points may be added or deleted (interior or exterior).
- Points can be dimensioned (interior), and all- or none-aligned.

Dimensions

End points must be dimensioned or aligned. Interior points may be dimensioned to reference geometry or to a coordinate system. Dimensioning to a coordinate system allows for manipulation of x,y positions via a table. This also allows for importing and exporting of ASCII text files that describe point positions.

Tangency

Tangency is the slope at which the spline will follow at the end point of the curve. Dimensions are applied as follows: (1) select the spline twice between the intermediate points; (2) select the end point at which you want tangency; (3) select a center line or reference plane; and (4) place the dimension with the middle mouse button.

Curvature

Curvature is the radius constraint at the end of the spline. Dimensions are applied as follows: (1) select the spline; (2) select the end point at which you want to control curvature; and (3) place the dimension with the middle mouse button.

✔ **NOTE:** *When creating a spline, start with two position points and then adjust the interior control poly points to shape the curve. Next, take the bottom-up approach in adding points only when necessary to assist in capturing a complex shape. The fewer the points on the spline the better for the following reasons: (1) The spline will be easier to manipulate and shape, and (2) Curve smoothness and acceleration consistency are more easily maintained.*

3D Splines

A three-dimensional spline is created by selecting various vertices or points within the model. This functionality is located in the **Datum ➥ Curve ➥ Thru Points** menu selection. Tangency constraints are optional at the beginning or added later through a feature redefinition. Tangency is applied at the end points to other splines, lines, axis, or normal to surfaces. Interior points ahve the option to be readjusted, added, or deleted. Straight line segments with interior radii is an optional selection that is useful when creating piping features.

Importing Point and Curve Data

Because Pro/ENGINEER treats imported curve data as one feature, it is important to extract only the required geometry. Extra data make the model inefficient and slow. Importing data is a very broad and complex issue. This section will focus on points and curves only as they relate to surface design.

Points

Points may be tagged and created with the native design model or imported from an outside ASCII text file. Prior to importing the point data, it is usually easier to first parse and edit the data using a commercially available spreadsheet program. Parameters to keep in mind when importing points follow:

- Imported points reference an offset coordinate system.
- Points may be read in with or without dimensions.
- The limit for a point file with dimensions is about 150.
- The limit for a point file without dimensions is about 15,000.

Curves

Depending on the available Pro/ENGINEER modules, wireframe data (curves) are imported as IGES data, datum curves from file, SCAN curves, or as Pro/LEGACY data.

IGES Data

Imported IGES curves allow for limited control over data manipulation. Individual curves may be deleted or changed; color one at a time or through a pick box. Arcs and circles translate as true entities when imported from IGES.

Datum Curves from File

This method reads both *.igs* and *.ibl* formats and offers much more flexibility in data manipulation when the feature is redefined.

✔ **NOTE:** *The .ibl files are written in a special format that Pro/ENGINEER recognizes. A simple program can be written to read point data and output in .ibl format. (See manuals for details.)*

IGES files read in as datum curves from file make much more functionality possible in terms of editing the data. All data are read in as splines so that true arcs or circles are not represented. Data editing functionality is summarized below.

- ❑ Editing—Editing of point positions that form the curve.
- ❑ Create—Generation of new curve entities within the feature model.
- ❑ Delete—Removing individual curve entities.
- ❑ Merge—Merging curves in one continuous curve.
- ❑ Smooth—Shaping curve through inputted factor.
- ❑ Sparse—Removing spline points within a given tolerance.
- ❑ Adjust—Commonizing multiple curve end points.

When surfacing to imported curves, always clean up curves before beginning the surfacing process. Once a feature is attached to the imported data, certain functionality is lost in the manipulation of the curves (e.g., merging and deleting become grayed out). Data sectioning is achieved by reading in multiple feature files and deleting unwanted curve entities.

SCAN Curves

Functionality for SCAN curves is similar to importing as datum curves from file. This option is available through the Pro/SCAN-TOOLS module interface. This method is used primarily for creating Class "I" surfaces from scan data of a physical model. Imported curves are automatically dimmed for visualization purposes. One advantage of using this method is that there is more functionality in creating smooth clean curves (Style) and checking distance deviations to surfaces.

Pro/LEGACY

This module allows data imports from other CAD systems as true arc, circles, splines, and lines with extensive functionality for manipulating and editing the data. This option is used for updating and editing existing legacy data.

✔ **NOTE:** *It is easier to refit through a network of curves than raw point data. When reading in scan data as an array of points, create a 3D curve through the point array. Pro/ENGINEER will pass a continuous curve piped through all these points. For more flexibility, IGES out and then back in as datum curves from files. This is a quick and easy way of generating curves from raw point data.*

Curve Analysis

Throughout the design process it is essential to evaluate various properties of design curves. Advanced analytical and graphical tools are located within the Crv Analysis menu. These tools measure arc length, tangency, radius, curvature, and deviation.

The curvature display graphical tool is the most important for simultaneously displaying both smoothness and curvature. Analytical tools are used to evaluate entities when hard empirical data are required. Use graphical tools early in the design process to help guide the design. Analytical tools should be used as final checks to confirm graphical evaluations. Curves should be evaluated at the time of creation to ensure quality.

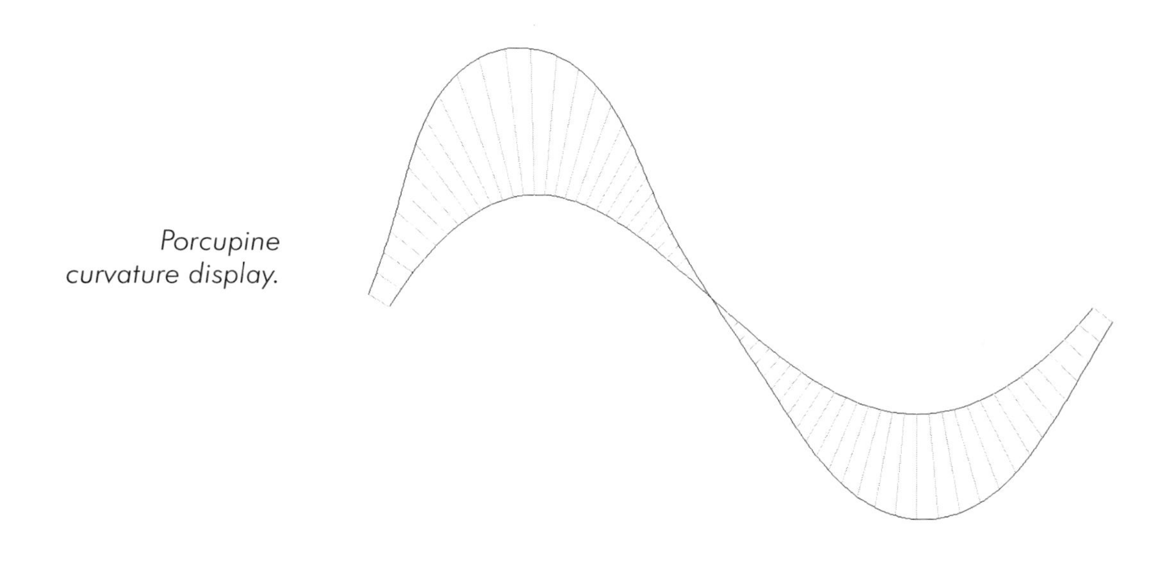

Porcupine curvature display.

Curvature Display

The Curvature Display (Crvture Disp) tool places equally spaced normal line segments that reflect curvature along the length of the curve. The length of the line segments is related to the slope of the curve at a particular position. Crvture Disp is the most frequently used and important tool for graphically evaluating a curve's smoothness and curvature. This curvature analysis tool is located in the Info Crv Analysis menu and the Sec Info menu within the Sketcher. Tips on how to read porcupine quill information follow:

- Longer vectors indicate more curvature, and shorter vectors indicate less curvature.
- Conics and two-point splines produce perfect accelerations.
- Scale porcupine as needed in order to fine tune smoothness.
- The smoothness of porcupine transition is extremely important.
- Smoothness data are written to a file called *curv#.dat* for further evaluation.
- The smoothness of a curve is also dependent upon the accuracy of the model.
- Class A curves and surfaces should be developed at a higher accuracy than the default standard (.0012). Use .0001 to .0003.

Conic Curvature Accelerations

Appearing in the next four illustrations are curvature displays of common conic RHO values.

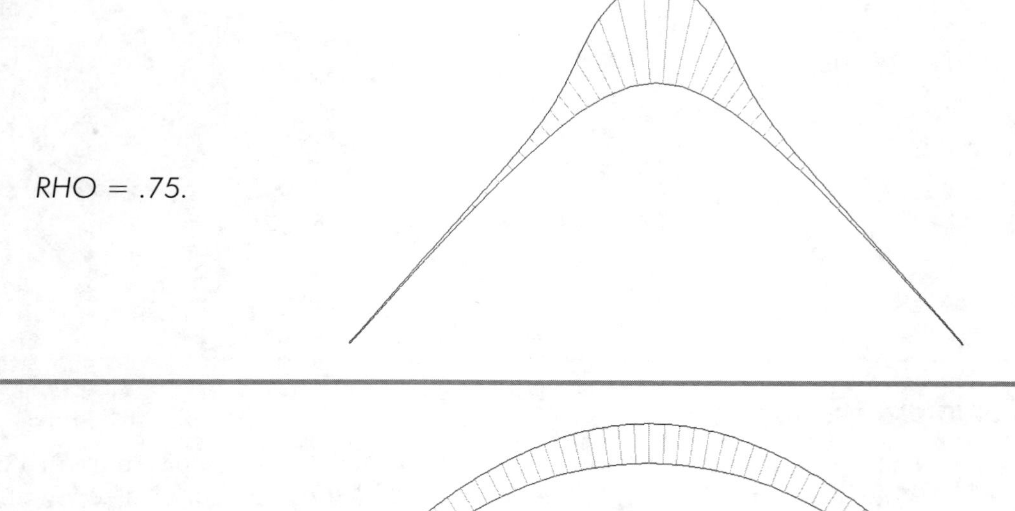

RHO = .75.

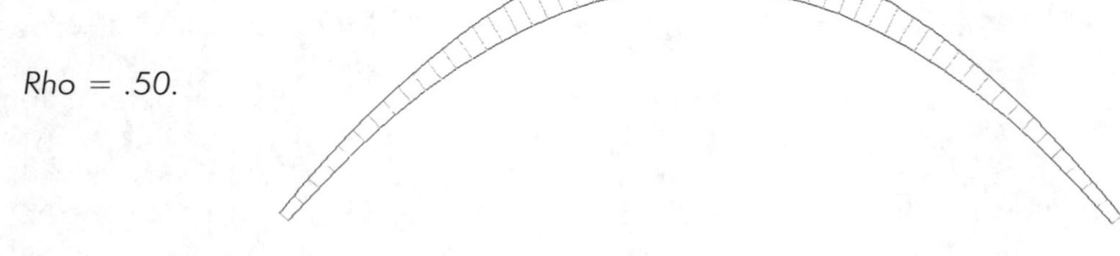

Rho = .50.

Rho = .414.

Rho = .25.

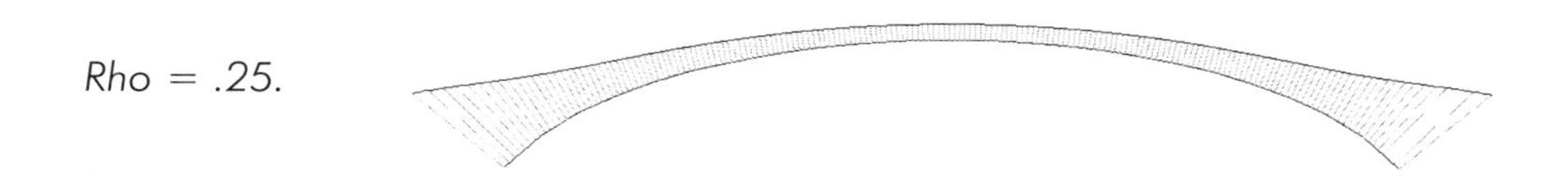

Spline Curvature Accelerations

Appearing in the next two illustrations are curvature displays of free-form splines.

Inconsistent accelerations and inflection points are undesirable.

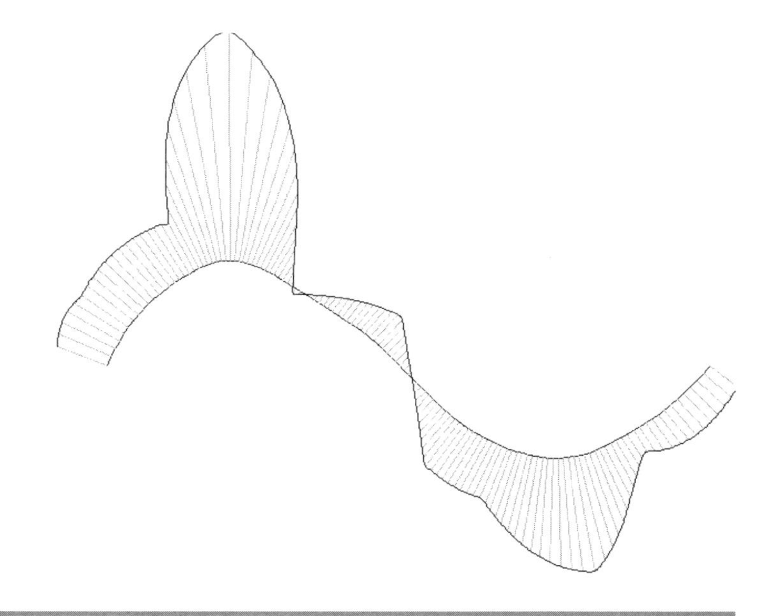

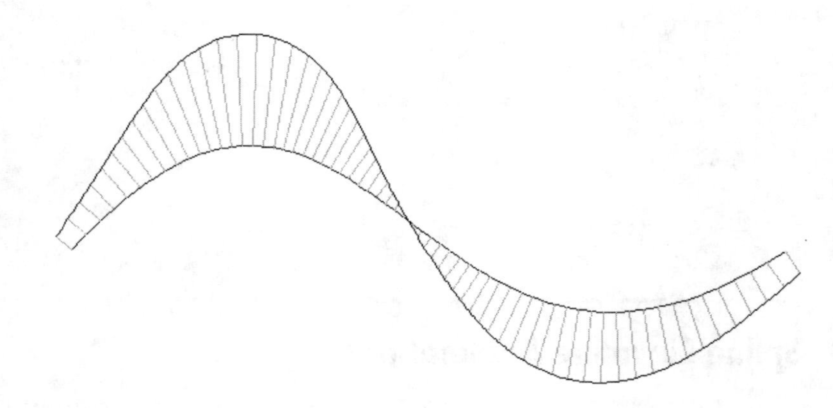

Clean, smooth accelerations with inflection point are desirable.

Example of Straight Line and Arc

The following example demonstrates both analytical and graphical checks of two curves at a common vertex.

1. Create a straight line and datum curve (arc) within one feature.
2. At the vertex, analytically measure the slope (curvature) at the end of each entity. In the Crv Analysis menu, select **Info → Measure → Curvature → End point**. The slopes should be the same.

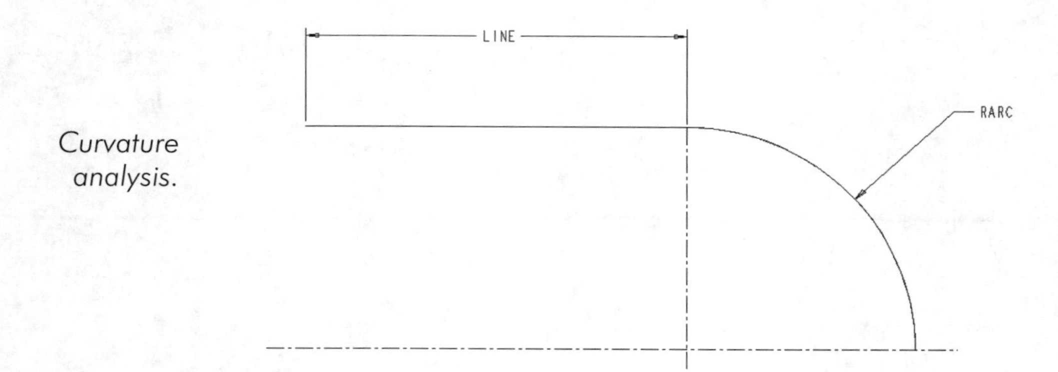

Curvature analysis.

3. Measure the tangency vector on the end point of each curve. Select **Info → Measure → Tangency → End point**. Tangency is indicated when the slopes of the two curves are the same.

End point slope vectors.

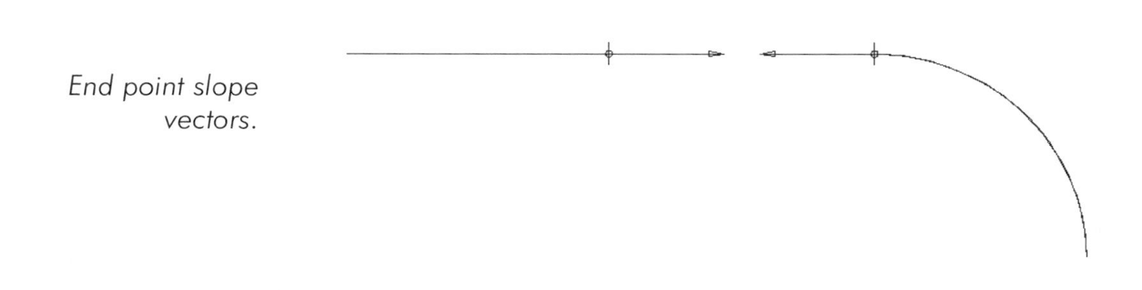

4. Measure the curvature of both curves through graphical and analytical analyses. The graphical check is located under the Crv Analysis menu. For the analytical check, select **Info → Measure → Curvature → End point**.

✔ **NOTE:** *The line has zero curvature, while the arc has curvature equal to 1/R.*

Graphical curvature acceleration check.

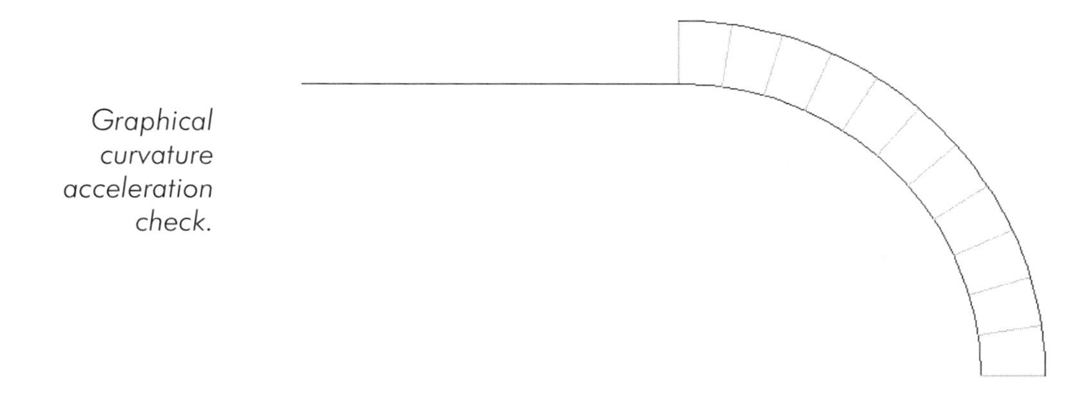

5. Construct a datum composite curve (approximate) through the two curves. This will create a C2 continuous curve that references the two underlying curves.

✔ **NOTE:** *C0 defined: End point positions are the same for two curves. C1 defined: Tangency between two curves at common point position. C2 defined: Continuous radius curvature blending between two curves at common point position.*

6. Measure the graphical curvature of the new C2 continuous curve.

✔ **NOTE:** *The results are a first-derivative continuous blend, with discontinuities in the second derivative. This is caused by blending a straight line, which has zero curvature, with a curve. The blending function causes an abrupt change in curvature leading to a dip in the new curve. This is highly undesirable because it will propagate back into the surfaces.*

✘ **TIP:** *Avoid straight line segments blending into curves! Instead of a straight line, use a semi-flat conic with the normal pointing outwards. This combination reflects the light better than a flat surface.*

Example of Multiple Curves

The following example demonstrates how curves blend into each other with both tangency and (C2) continuity.

1. Create three independent planar curves tied together at two common vertices points (conics on the outside and a two-point spline on the inside).
2. Within the Sketcher, hold tangency constraints on the middle curve as it blends into the adjoining curves.

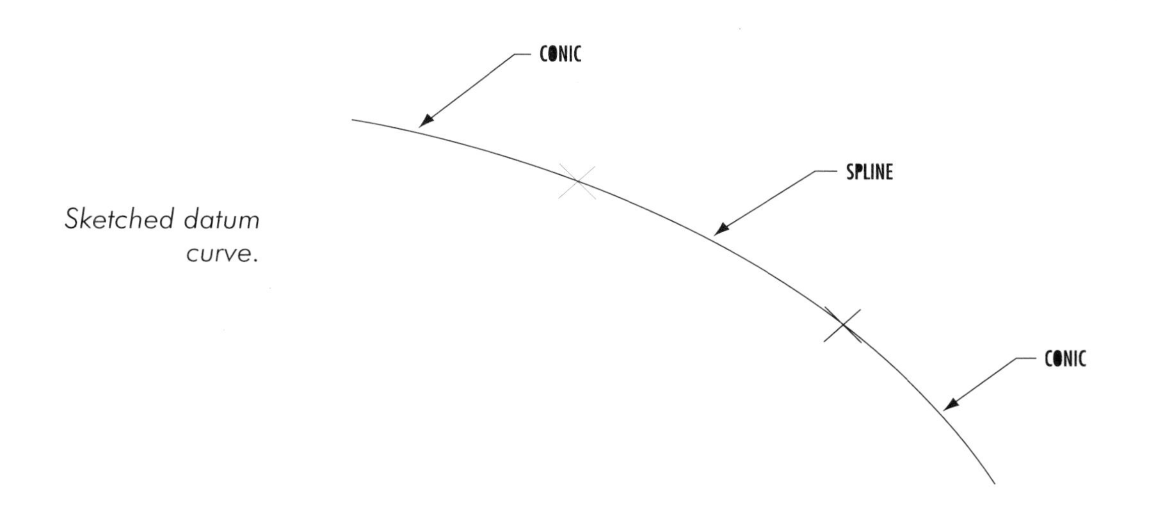

Sketched datum curve.

3. Measure the curvature accelerations using the Crvture Disp option (Crv Analysis tool).

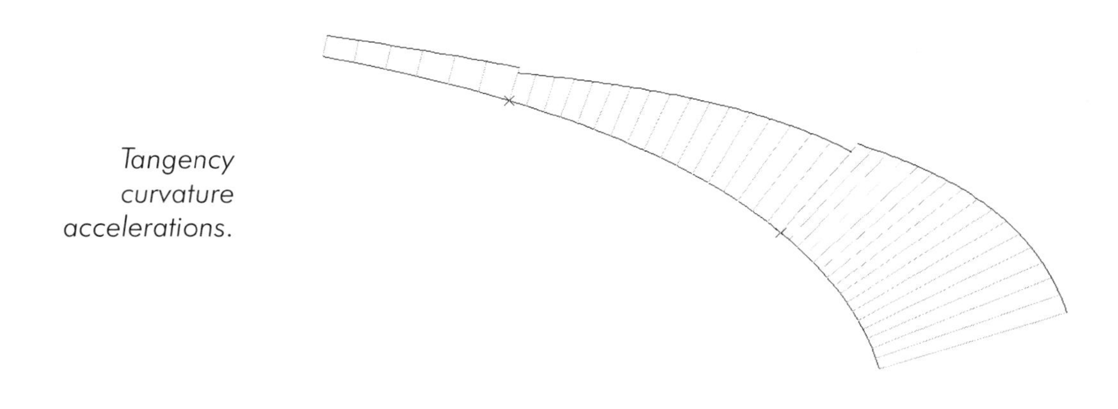

Tangency curvature accelerations.

✔ **NOTE:** *At the vertex points the quills do not line up as they flow from one quill to another. The curves have tangency constraints but are not C2 continuous. In other words, the curvature (1/R) at the end of each curve is different.*

4. Analytically measure the tangency and curvature at the end point of each curve using the following picks: **Info → Measure → Curve-Edge → Tangency-Curvature**.

✔ **NOTE:** *The tangency vectors (X, Y, Z) at each vertex should be the same, but point in opposite directions. The curvature at the end point of each curve should be different.*

5. Measure the radius of the end points of the two outside curves at the vertice where the middle curves connect. Record these values.
6. Redefine the middle curve and add a radius value to the end of the spline. Make the radius value common with the radius of the adjacent curve, and then regenerate.
7. Graphically measure the curvature accelerations of all three curves. The accelerations should flow through at the common vertex points and appear similar to the following illustration.

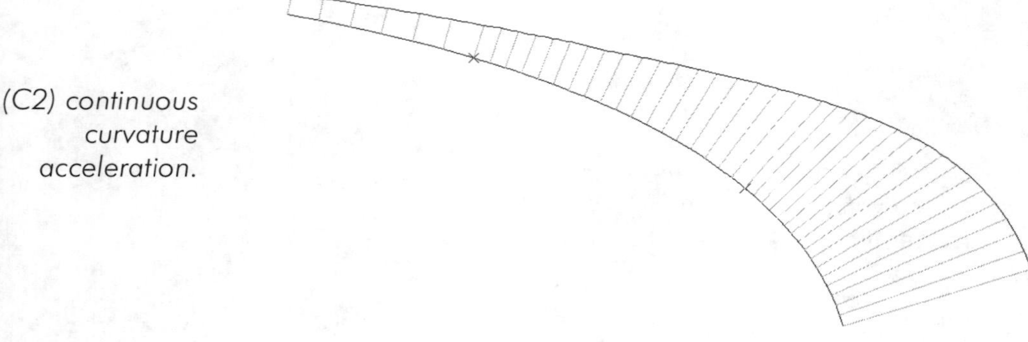

(C2) continuous curvature acceleration.

The disadvantage of adding a radius value to the end of the spline is that the spline becomes unmodifiable. If you do not have enough points to capture the original shape and then lock in the radius value, this function will add little value. A continuous blend will be achieved, but 99% of the time the curve will have flat spots and dips in the middle. Because the curve becomes locked out, adding a radius value is a seldom used function. The work-around for is to add a connecting middle curve using the Pro/SCAN-TOOLS module.

Quick Fix

1. Quit window, access the Pro/SCAN-TOOLS module and recall the part.
2. Create a style curve through the two end points (Pick points option) and then modify.
3. Access the Setup display and query select the style curve in which you want to highlight the curvature.
4. Select **End Point ➡ MkContCrvtr**. Select the style curve near the end points and then select the adjoining curves.

✔ **NOTE:** *Pro/ENGINEER will automatically adjust the curve to be continuous along with the accelerations.*

5. Adjust the curve by selecting Control Poly and keep the fixed menu. Lock out curvature for both end and beginning to modify the spline by selecting Move Point. Pro/ENGINEER will lock out two nodes deep (C2) within the curve and allow the manipulation of interior points. The curve accelerations update in real time as the desired smoothness and shape are achieved.

The disadvantage to the above process is that the style curve has no associativity to the original curves. If the original curves are parametrically moved, the style curve remains in the same place. This will cause the surfaces created through these curves to fail. The surface may be rerouted to a new style curve, or the original style curve may be adjusted to the new position.

It is much easier to create a new style curve and reroute the surface than adjust the original style curve. Planar curves seem to adjust well but three-dimensional style curves adjust with difficulty. It is much faster to create a new curve and reposition the accelerations. If you create a new style curve within the same feature, the ability to delete the curve is lost. Add it to a junk layer and blank off.

Surfaces

Surfaces offer flexibility in creating complex shapes. There are many ways to create surfaces within Pro/ENGINEER functionalities. We focus on creating surfaces from boundaries because it is assumed that most users are familiar with creating extruded, revolved, sweeped, and blended surfaces. Surfaces created from boundary curves offer the most flexibility in creating and manipulating complex shapes. You have total control over the boundary shape and conditions that control the surface. Curves, tangency, and control points are key items controlling the surface. The quality of the surface is dependent upon the quality of the entities selected to form the surface. It is important to understand that surfaces are created by blending reference geometry in two directions (u,v). Thus, control of the direction and magnitude of the underlying geometry in both directions is essential. The quality of the surface is dependent on the quality of the curves. Curves, control points, and tangency are key items that control the surface when creating a boundary blend.

Curves

Surfaces can be created from curves, edges of surfaces, edges of solids, or points. When entities are selected, they are tagged and stored for adding, deleting, or editing. Surfaces may be created by blending in either one or two directions. Using the chain command allows for more flexibility in trimming the entity for the bounding blend.

> ✔ **NOTE:** *Blended surfaces from curves in one direction result in less control over the tangency influence of the surface. Curves in the other direction will help control the magnitude of the blend. Do not assume that the software always generates a good blend. Controlling the surface in two directions maximizes your control over the blend of the surface.*

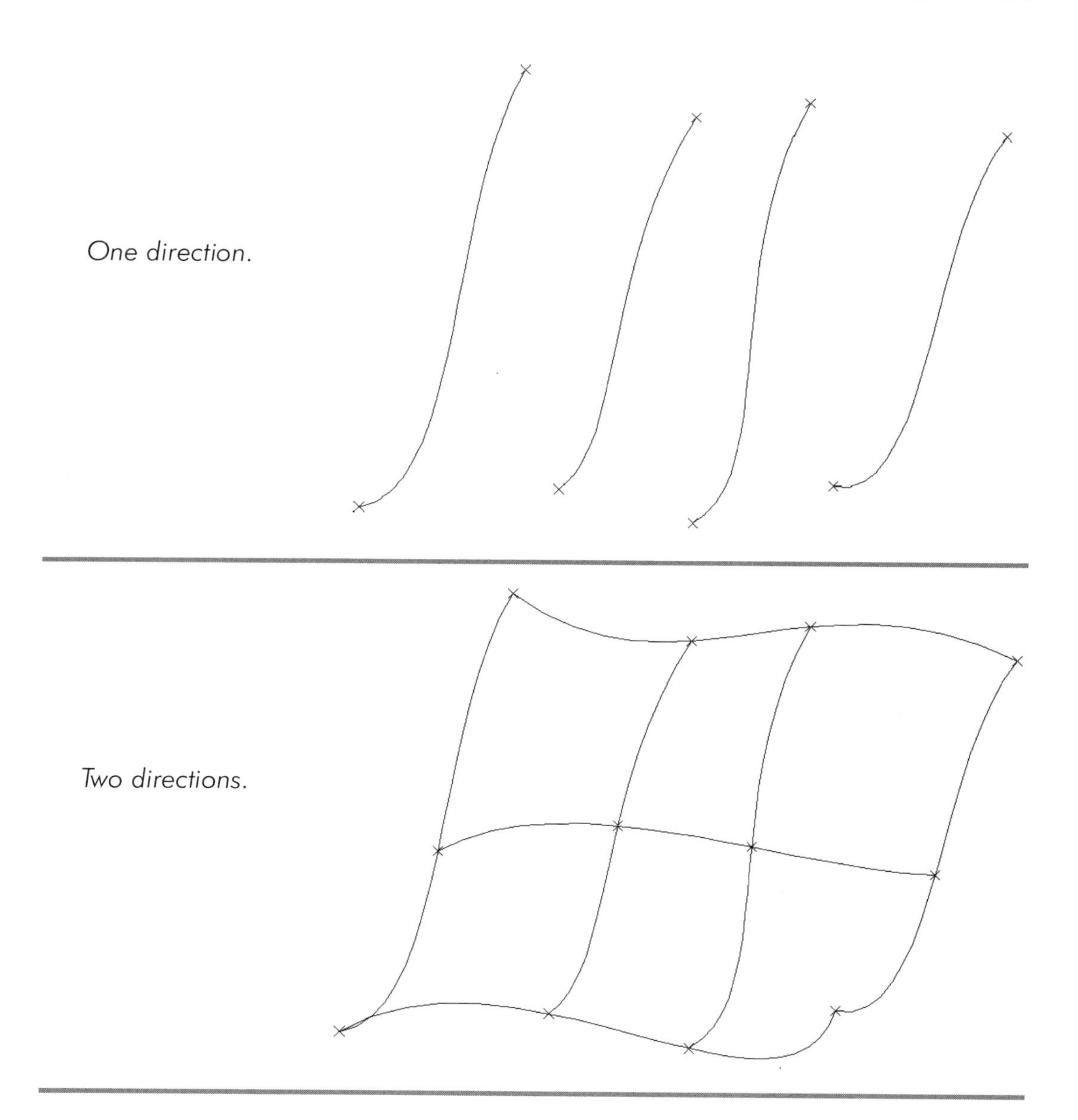

Control Points

Control points allow for more control of surface blending. Pro/ENGINEER will try and blend surfaces by ratioing the vertex splits between curves. In the case of intermediate break points in a curve, a surface created from such curve will reflect the same split. Sometimes surfaces need to blend a different way between curves. Pro/ENGINEER provides the ability to normally pick the blend points between curves which helps eliminate unwanted breaks in the surface blend.

✔ **NOTE:** *The goal in body design is to try and keep the number of patches to a minimum. When you quilt a surface with multiple patches, the breaks automatically propagate back to the new surface. Try and keep the patches to a minimum without sacrificing the fidelity of the surface.*

Tangency

This option provides you with the ability to blend the entire edge of a surface tangent with another surface. The curves used in generating the surface must be tangent in order to have tangency on the surface. Tangency may be added or removed at any time in the design process. TangInedge automatically assigns the tangent patch surface with the appropriate blended edge, or you select the appropriate patch.

Tips for Creating Boundary Blended Surfaces

Tips and suggestions for creating boundary blended surfaces appear below.

- ❑ If a surface will not generate between a set of curves, take the following steps: (1) Randomly skip curves to isolate the potential problem area. (2) Create as free boundaries, and then refine and add tangency constraints. (3) Try creating curves in other orders until surface generates because the u,v direction affects how the blending of the surface occurs.
- ❑ Non-tangent curves are the main cause of surfaces failing because curves must be tangent in order for surfaces to have tangency conditions.
- ❑ The C2 pick only works if tangency is first applied and the reference curves have the same radius of curvature at the end points. C2 blending is only possible in one direction.
- ❑ Use the Review button to view the flow of the blending lines before the surface is accepted.

Surface Analysis

Surface analysis tools include porcupine sectional curvature display, curve deviation, inflection points, and linear and logarithmic guassian color-shaded display. These tools help evaluate the quality of the surface in the design process. The graphical color display tools generally provide directional information while numerical data is obtained for final documentation. These tools are located in the **Info → Srf Analysis** environment. The most frequent and valuable tools used are the Max Dihedral and surface Porcupine options.

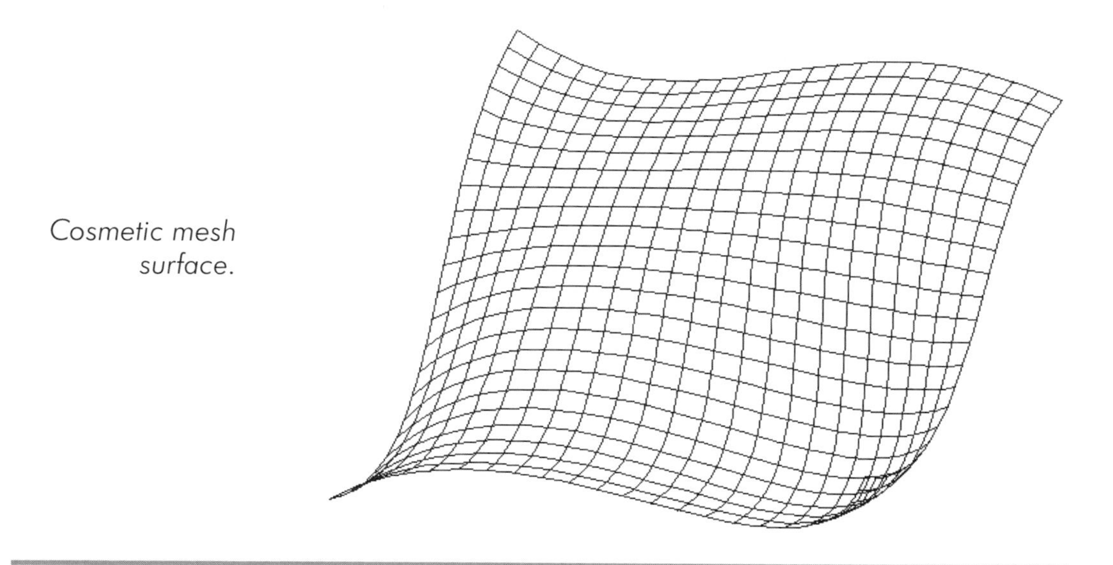

Cosmetic mesh surface.

Max Dihedral

Max Dihedral checks for the maximum tangency angle between two surfaces that occupy the same boundary. A few pointers follow:

- Surfaces must be quilted in order to check tangency.
- The accepted maximum angle for class "A" surfaces is < 0.1 degree.
- If the angle is 0.1 degree or higher, and the surface was created with tangency constraints, add another control curve around the area of maximum angle, redefine the surface, and select new curves.

Porcupine Smoothness

Porcupine is the most valuable tool for reading the curvature display of surface. It is similar to Crvture Disp but displays the entire surface in two directions (u,v). Consider the following when using this tool:

- Scale and density control are subjective.
- Attempt to view quills in true view to understand the flow of the surface.
- Adjust the surface according to common local accelerations.

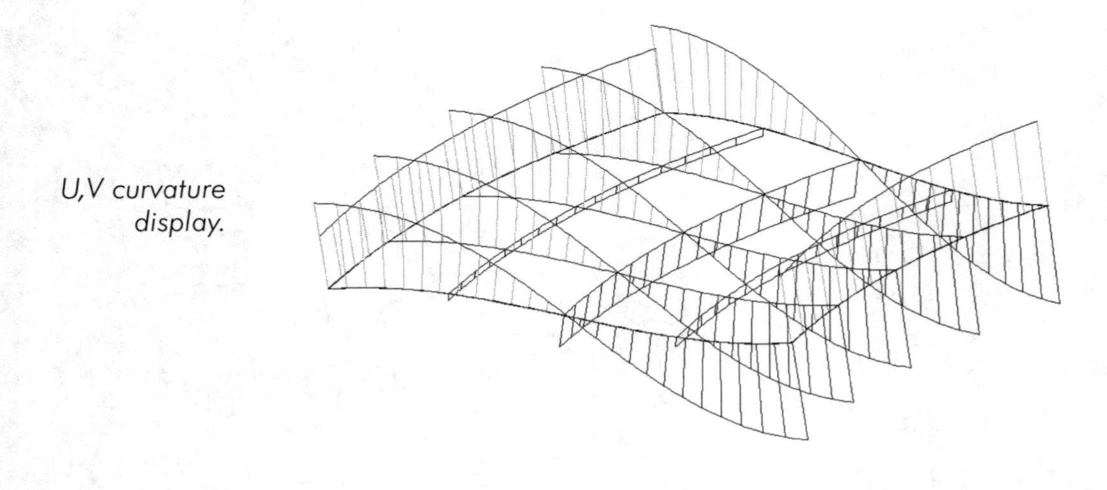

U,V curvature display.

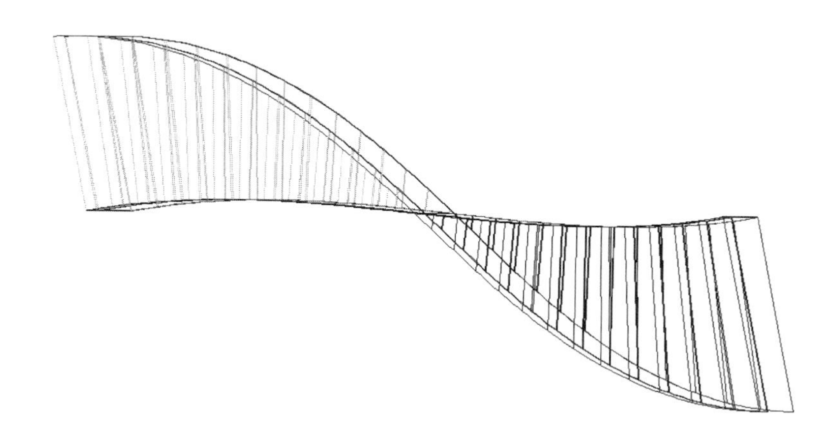

True view evaluation.

Color Rendering Analysis

Shaded color spectrum graphs are displayed to demonstrate curvature analysis as guassian or section slope evaluations. The plotting grid may be displayed as linear, logarithmic, or two-color with additional control over upper and lower boundary limits.

Color analysis averages results with multiple patches. Thus, you should display individual surfaces for analysis by blanking out surrounding geometry. This will yield a more accurate analysis of the surface when using a rendering analysis. These tools are located in the Surface Analysis menu.

✔ **NOTE:** *Color spectrum analysis is used for identifying change in surface curvature.*

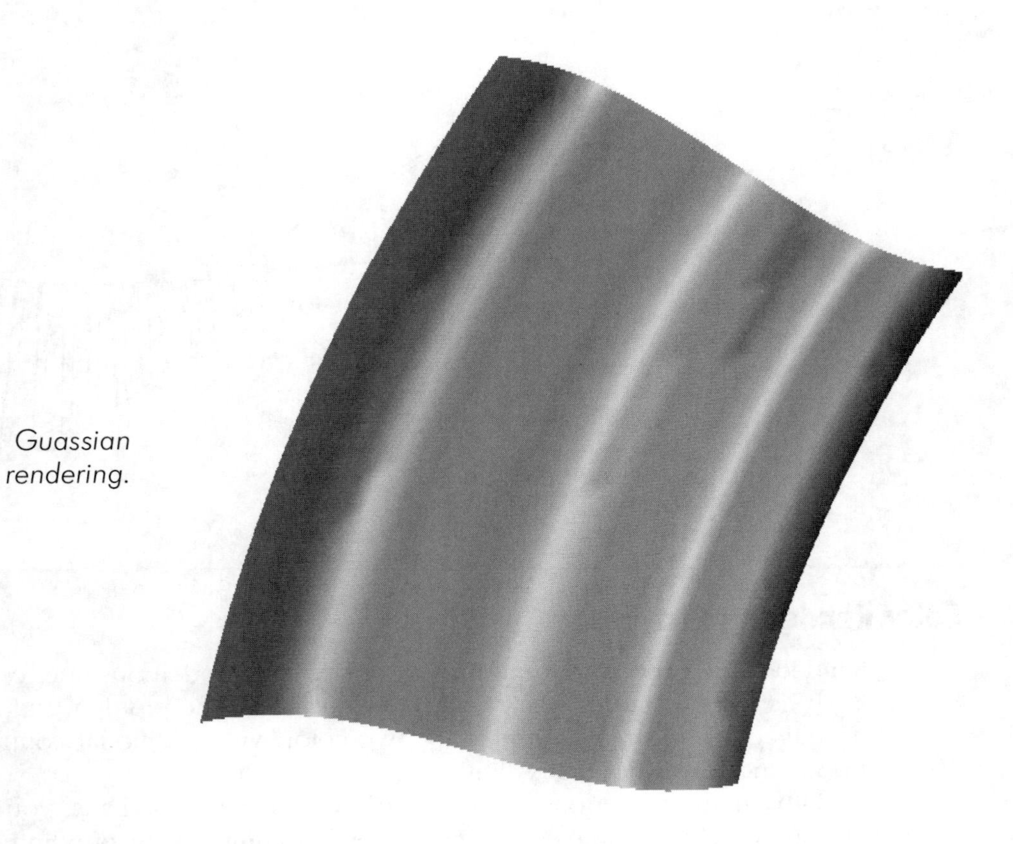

Guassian rendering.

Reflect Curves

This tool allows for setup of tubular light sources to display multiple reflection lines on a surface. Only one direction with one or multiple lights may be displayed during analysis. This tool is located in the Surface Analysis menu.

✔ **NOTE:** *Cutting multiple sections in the x,y,z plane produces more desirable results. Always view the sections in a normal 2D view to get a topographical perspective. Visually and analytically check to see if section curves are smooth.*

The following steps are recommended when cutting multiple sections:

1. Create an offset datum plane to the x, y, or z master plane. Create a datum curve by intersecting surfaces.
2. Create a local group of the section and plane and then pattern to desired number of sections and spacing.
3. Blank out surfaces and begin to measure tangency and curvature end points to determine how well the surfaces blend into each other.

Summary

Analytic and graphic tools play a major role in displaying feedback on surface quality. This information is subjective, but assists in making engineering decisions on how to manipulate the surface. Time and experience are necessary for feeling comfortable with evaluating and changing the data. As a rule of thumb, rely on graphical checks early in the design process, but always follow up with analytical checks before going on to tooling.

Starting a Surface Design

Introduction

This chapter focuses on laying out the blueprint for initiating a surface design project. Setting up a model for free-form surface design consists of model management, slab theory, and framing structure. Starting an open surface model differs from designing with solids in that a more structured environment is required for sectioning off strategic areas of the model.

Whether working with scan data or simply developing concepts, two basic rules apply in starting a surface design. First, you should start with the basic shape of the model and add more detail as the design develops. Next, the basic shape must be comprised of global slab surfaces that form the bulk shape of the design. Intermediate or blended surfaces tie the slab surfaces together to form the shape of the design. This procedure allows the design to be more flexible and efficient for maintaining surface quality after multiple changes.

Setup

When beginning to model either conceptually or with scan data, it is essential to start the model in a logical and organized manner. Selected key steps for organizing the model for maximum flexibility are discussed below.

Default Datums

Start with default datums and a coordinate system. Rename each datum plane and coordinate system to reflect the zero datums for X,Y, and Z. This is an essential aid for selecting these features by menu and tying all the reference curves to common planes. In addition, the procedure helps to manage undesirable parent/child relationships.

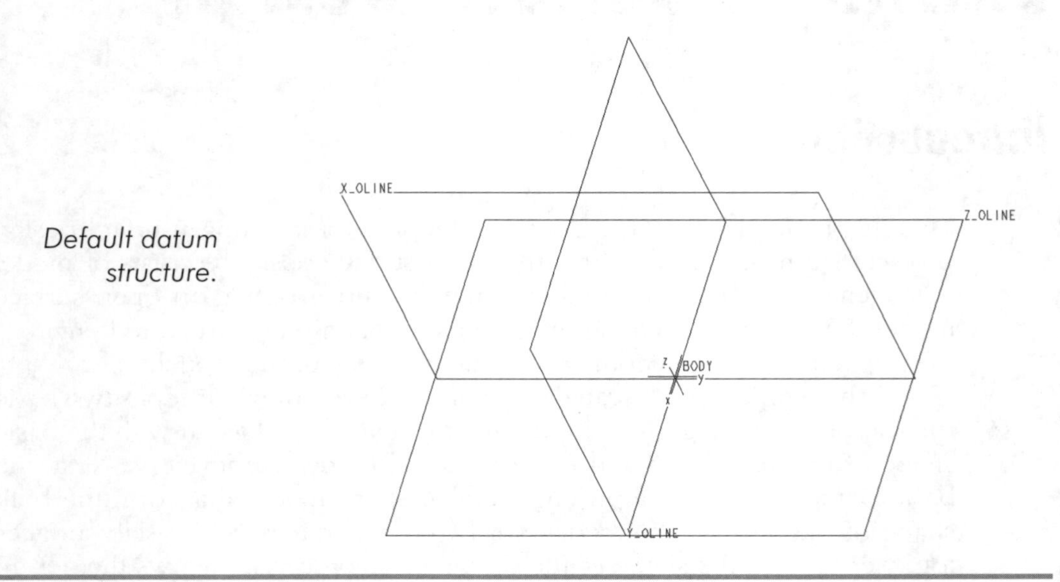

Default datum structure.

Importing Data

Imported data are received in a variety of forms such as clay scan data, raw ASCII points, or IGES curves. If you are not using the Scan Tools module, always import wireframe or scan curves as datum curves from a file. Importing curves through IGES limits your ability to manipulate the data. Clay scan data will usually be zeroed and scanned in the x, y, and z directions.

The imported data should be parallel and perpendicular to the default planes. If the data is not orthogonal, create an offset coordinate system with six degrees of freedom. Manipulate the offset coordinate system until the desired location is achieved. For clay scan data the spacing between each grid is subjective to the size and shape of the part. Flat shapes with little curvature require less scan definition than areas that have high curvature shapes.

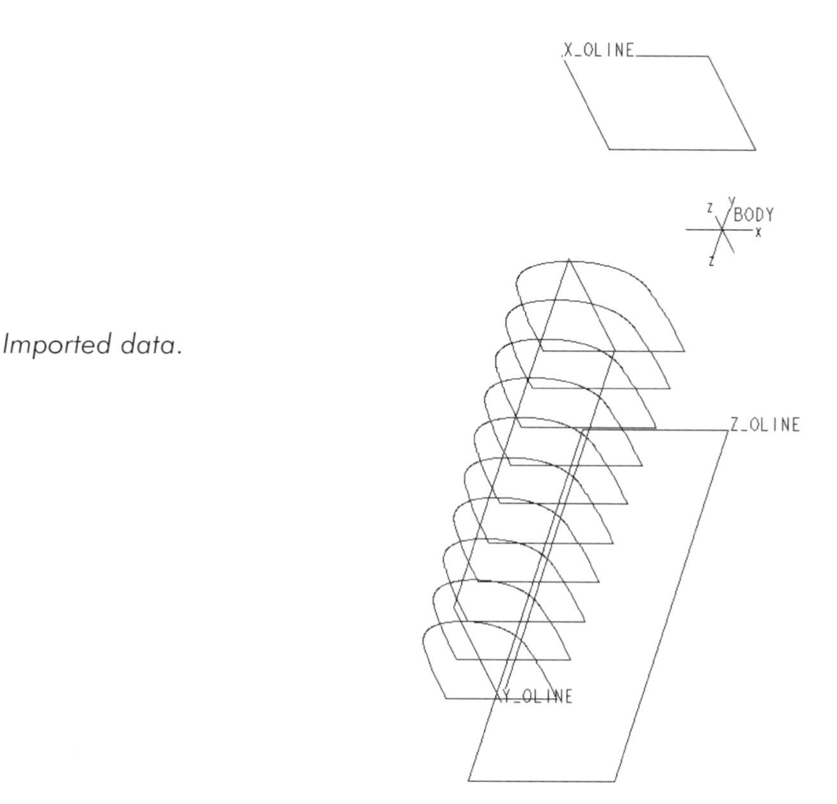

Imported data.

Goals for Working with Complex Shapes

The major goal in working with complex shapes is to break down the data into simple shapes. To accomplish this, model structure and management are key for simplifying the geometry. A useful breakdown method is sectioning off the data in such a manner that only the data portions required at the time of new geometry construction are displayed.

✔ **NOTE:** *A common misconception when working with scan data is that the design needs to reference every section. This practice presents two problems: (1) Working with every available section is not practical nor efficient; and (2) Creating smooth curves and surfaces through such a multitude of reference points becomes virtually impossible.*

Working sections need to be targeted with the spacing between the sections subjective to the designer. As a rule of thumb, you need a finer spacing of key sections when capturing shapes with higher rates of curvature compared to relatively flat planar areas.

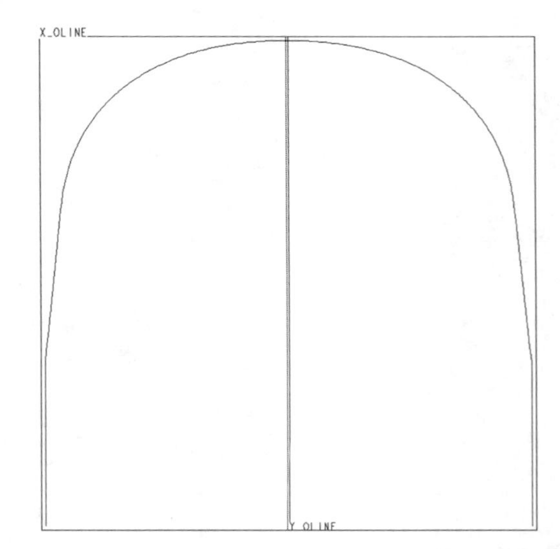

Isolated section.

Scan data are comprised of 2D cross sections related to the orthogonal default datum planes. The goal is to be able to isolate and display key sections while working sections are developed. This is accomplished through a layering scheme that relates the local key sections to corresponding named datum planes and layering conventions.

Create datum planes parallel to the 0 lines at desired curve locations. Rename the planes to reflect their positions. For instance, a plane created 500mm from the X_Oline would be labeled *X500*.

Create a layer of the same name (e.g., *X500*), and attach the datum and geometry to the layer. Creating layers in this fashion allows for the isolation of a particular section while a working section is being developed.

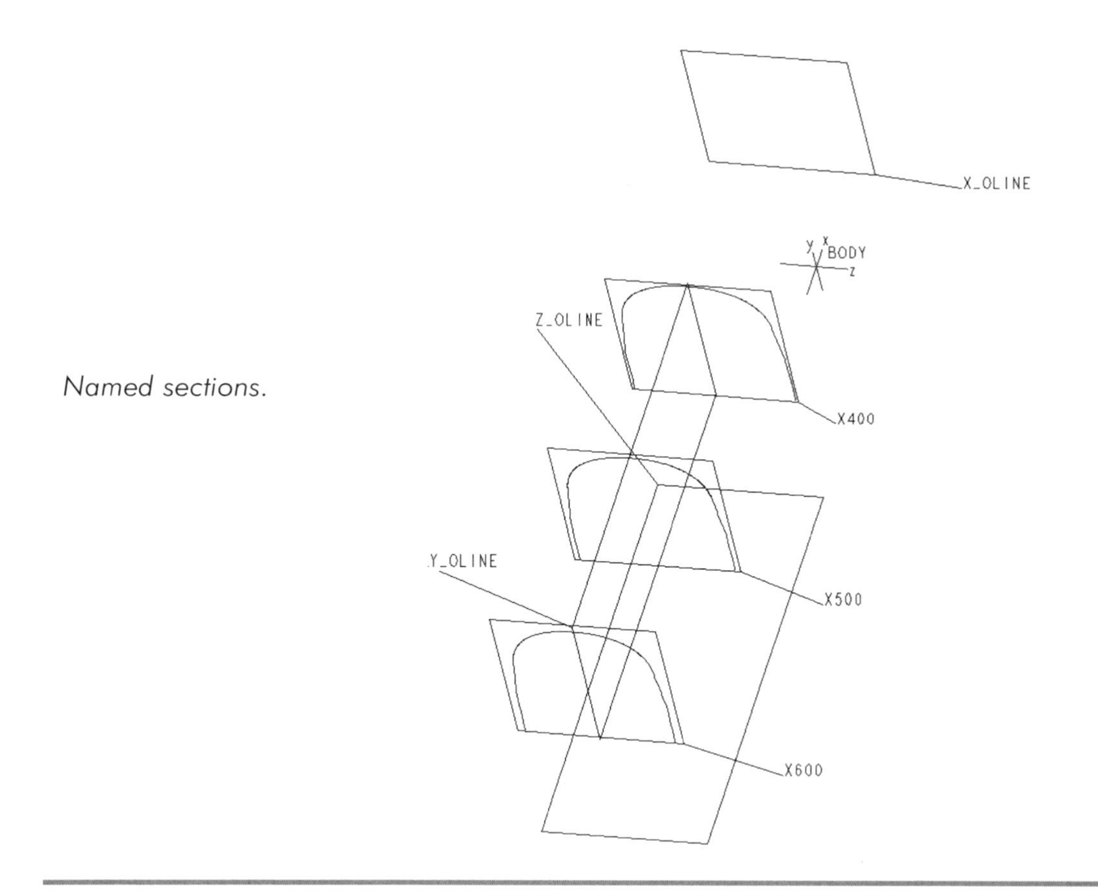

Named sections.

The above procedure permits increased model management for selection by menu, isolating lines, suppressing, view display, and x sections. The key is to keep it simple by displaying one section at a time in the development process. Layering and model management are fundamental to surfacing and any part development process.

Framing

The process of framing a section further explores the idea of breaking down the design into basic shapes. Framing or "coach building" is well known, and basically captures a complex cross section through easier, more manageable shapes. Overbuilt or slab curves are created that capture the outline of the cross section. These slab curves are used section by section to help shape a slab surface.

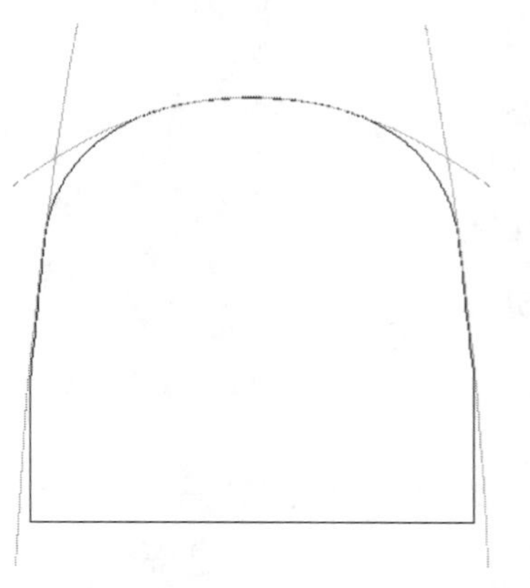

Framed section.

Tips for framing a design follow:

- Break up 2D cross sections into simple slab curves (conics and splines). When creating conics and splines, eyeball the curve on top of scan curves. Measure the deviation later. It is important to lay out the curve at this point. The curve will move many times until it finds its final position. Copy curves when possible to save time.
- Try and create each curve as its own feature. This practice builds more flexibility into the model for potential downstream changes.
- "Commonize" 2D end points at constant distances from the reference planes. This allows a planar curve to be generated in the other direction.

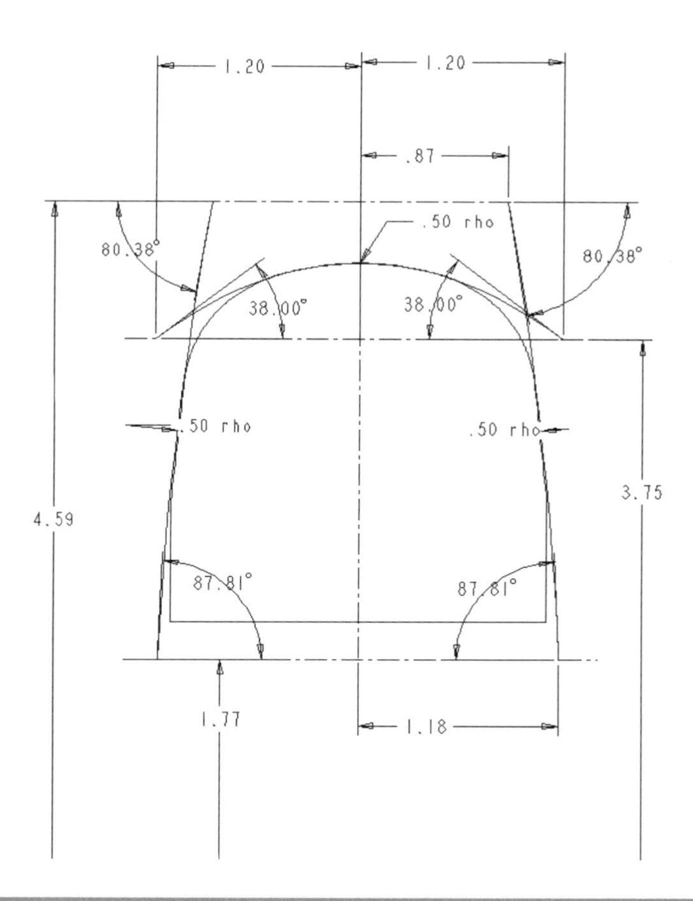

Creating a curve as its own feature.

- Begin to frame only key cross sections based on the shape of the surface The distance between sections is dependent on the acceleration of the surfaces. Relatively flat planar sections need less definition than high accelerated areas. Again, not every scan section should be used in creating a new model because such practice would be very impractical. It takes time and experience before you become comfortable with selecting key sections.

✔ **NOTE:** *Surfacing is quite subjective because each person will pick a different set of curves to work with. Thus, the art of surfacing is based on the fact that every individual will make different decisions based upon how s/he interprets the design.*

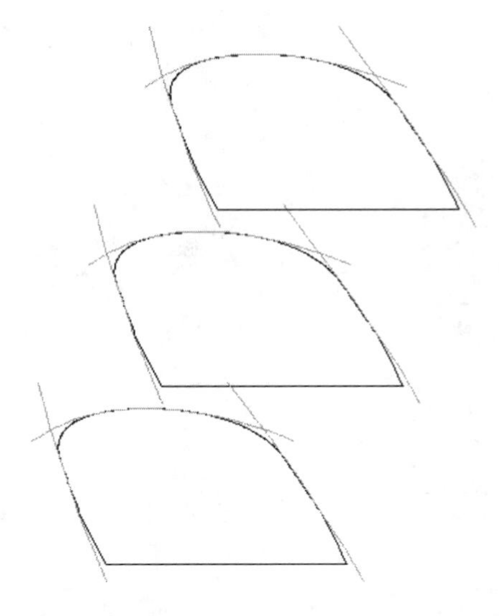

Multiple framed sections.

- ❑ Change the color of the newly sketched curve to distinguish it from the scan data.
- ❑ Place datum points on the ends of the vertex to help align conics or splines when sketching in 2D.
- ❑ Create addition points by intersecting planes with curves.
- ❑ Create the plane offset to lines for flexibility. Rename or label and place on layer.
- ❑ Use the plane to sketch a new curve in the opposite direction. Make sure that you align to the predefined datum points. Use tangency constraints with angular dimensions on the curves created in the opposite direction. This procedure helps in fine tuning the surface by controlling the slope in two directions at the corner of a new surface.

Example of Creating A Blended Surface

The following example demonstrates the technique of creating a blended surface between two slab surfaces. Gaining complete control of the direction and magnitude of the blend is the theoretical base for constructing a surface in this manner. The surface tangent line is the key to how highlight lines react at the edge of contact.

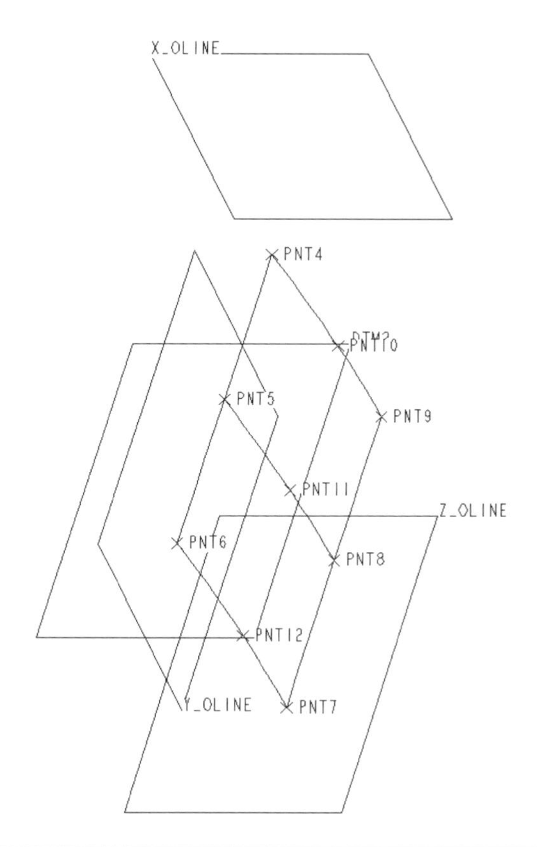

Wireframe slab.

Procedures for creating a boundary blended surface follow:

1. Create slab surfaces that are overbuilt and intersect one another. These surfaces can reference a framing structure from key master sections or created as free-form sweep blends. Make sure that the surfaces fully intersect each other.

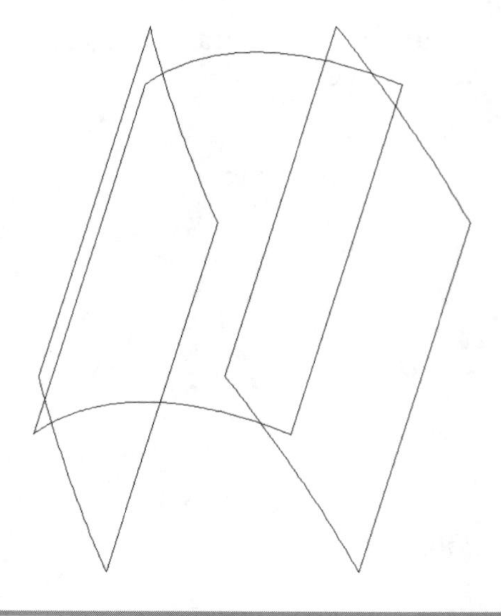

Slab surfaces.

2. Visually check the surface intersections by shading the model and visually checking through dynamic rotation.

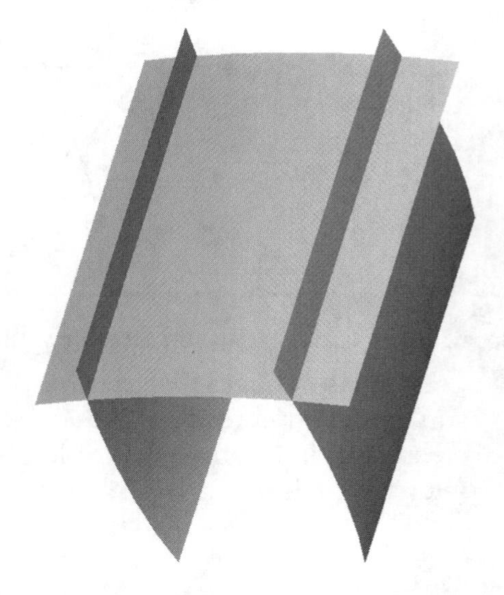

Shaded surface intersection.

3. Perform an analytical check of the surface intersections by creating a datum curve from the intersection of two surfaces.

✔ **NOTE:** *In body surface design the purpose of the theoretical intersection is to provide a reference line for creating the blended surfaces tangency lines. The tangent lines dropped on the slab surfaces should basically follow the shape of the theoretical intersection for a constant blend. This promotes a cleaner blend between the two intersecting surfaces upon visually checking the highlight lines. If an uneven tangent line is created, the highlight line will wiggle as light passes over the common boundary.*

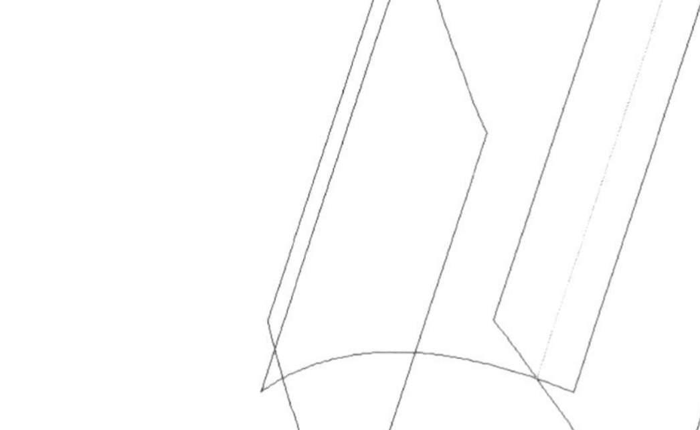

Analytical surface intersection.

✔ **NOTE:** *Creating a datum curve through intersecting surfaces also helps when two surfaces will not quilt together. If two surfaces will not merge or intersect, try creating an intersection curve between the two surfaces. This process will usually isolate the problem area, such as a gap in the curves indicating that the surfaces do not fully intersect.*

4. Check the slab surfaces independently for smoothness or deviation from scan data before the blended surface is created. A bump or blister near an edge of a slab surface will propagate back into a blended surface.

5. Once the slab surfaces are checked for accuracy and smoothness, begin the layout where the blended surface begins and ends from one slab surface to another.

6. At the end of each patch (slab) surface, place a datum point (length ratio) where you think the blend should begin and end.
7. At each point, create a datum axis point normal to plane that is used for referencing the tangent blend curve. Create the axis from an on-the-fly datum plane that passes through the point normal to the curve. The axis will always be a tangent vector at any point position along the curve.
8. Create a blended curve on the common 2D plane at each end of the slab surfaces. Use the predefined points and axis to act as references for tagging the position and tangency of the conic or two-point spline. The curve blend points now master the relationship between the two adjoining curves. As the points are moved on the curve, the blended curve will move to the new position and keep a tangent connection. Now this is parametric surface modeling!

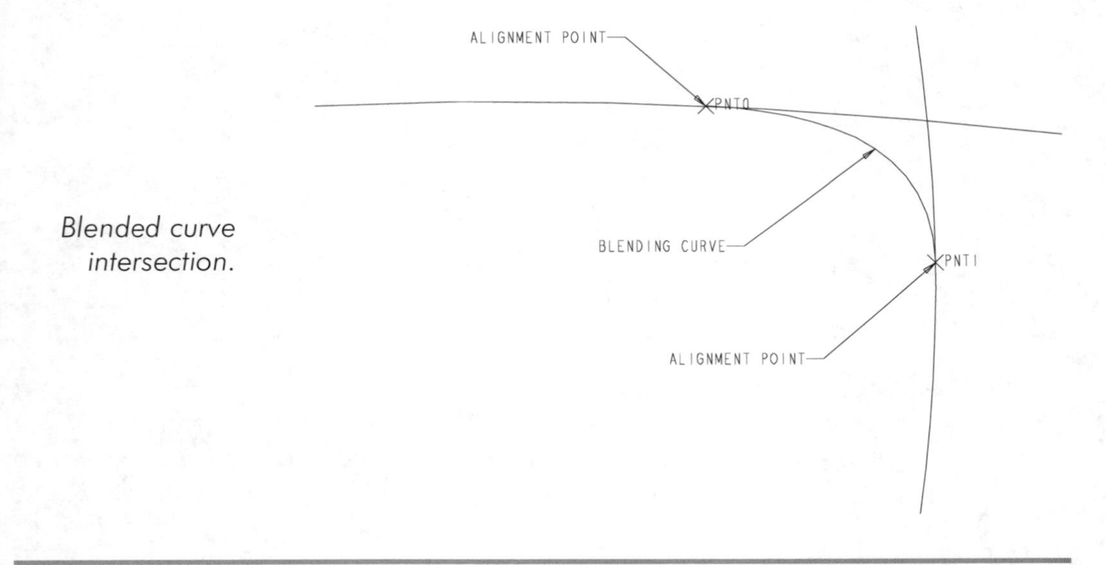

Blended curve intersection.

9. On offset datum planes, create datum curves (conics or splines) that represent the tangent line for the respective blended surface. Use the theoretical intersection line to help guide the flow of the tangent line. Verify that the curves are aligned to the predefined datum points which control the location of the blend. Evaluate the planar curves for smoothness.

✔ **NOTE:** *When blending tangent surfaces together, as a rule of thumb avoid exceeding a curvature ratio blend of more than 5 to 1. In other words, the measured curvature at the common blend point should not exceed five times the curvature of the slab curve.*

2D tangent lines.

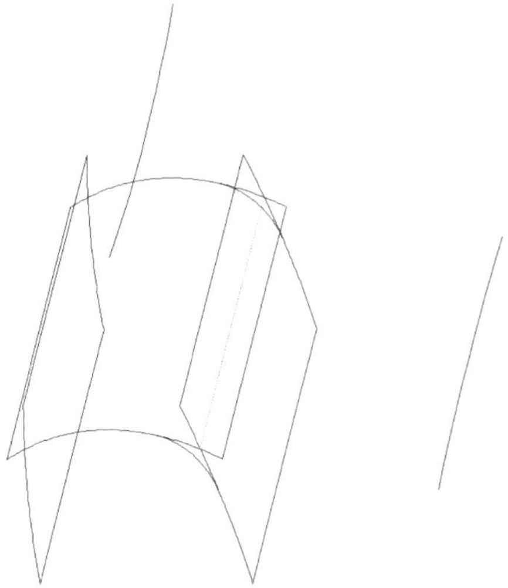

---

10. Drop these planar curves on the surface by projecting parallel to the master datum planes. This will create a four-boundary patch used for creating an intermediate surface.

11. Create a blended surface from boundaries with optional tangency. The direction and order of the curve selection affects the blending of the surface in the U,V directions. Test various directions to determine the optimal blend. Remember, if the construction curves are not tangent, the boundary surface will not be allowed to hold tangency.

12. It the surfaces together and begin evaluating the quality of the blend by using graphical and analytical checks. Typically, a graphical check such as cosmetic shade, or a rendering shape such as guassian or slope, is performed first.

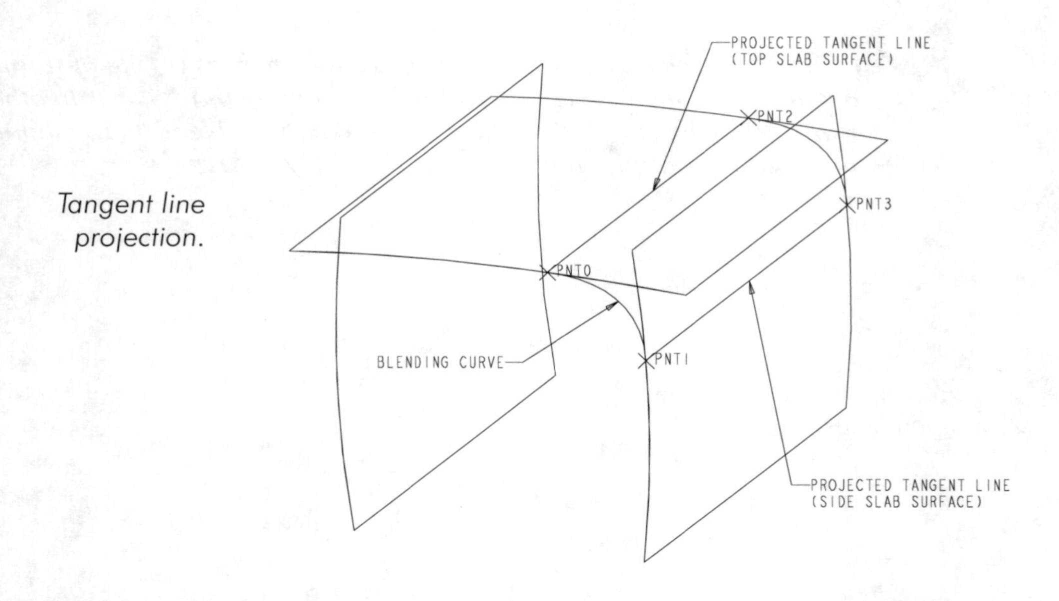

Tangent line projection.

! **WARNING:** *Do not rely on graphical checks to check the surface quality. Use analytical data such as max dihedral and cutting curves through the surfaces to fully evaluate the quality of the blend. (See Chapter 10 for information on these techniques.)*

✗ **TIP:** *When quilting the surfaces never quilt the intersecting surfaces before the blended surface. This sequence causes Pro/Engineer to trim two edges at the same time, which occasionally causes problems depending on the complexity of the surface. Always quilt the blended patch to one or the other slab first. This produces much better results.*

13. If additional surface control is required, add an intermediate curve between the two rounded sections. Insert a plane and intersect it with the two slab surfaces. Create a datum point and axis at the intersection of the new plane and the tangent line dropped on the surface. Add an intermediate curve that is tangent to the new curves. Redefine the surface to add the new curve in the blending direction. Adjust accordingly.

Summary

The creation of slab framing curves and surfaces is the key to capturing the complete design intent. Fundamental geometry skills are required to visualize and build the design through a network of parametrically designed construction features. The surface model is the end product of the construction features. Remember to keep the design simple, and add complexity as the model matures.

The Surface Design Project

Introduction

When beginning a surface design project, many factors will determine success. A surface design may be originated in Pro/ENGINEER as a native conceptual design or from reference data such as IGES or point data. Regardless of the way in which a design is constructed the basic fundamental rules of slab theory and patch structure are essential. This chapter integrates the design philosophies of Chapters 10 and 11 in approaching a design problem by breaking down the shape into simpler forms, and analyzing the integrity of the curves and surfaces.

Creating a Fully Parametric Base Cap Surface

The following example is a tutorial for creating a fully parametric class "A" surface of a base cap. IGES data, which serve as a background reference for the new design, are imported. The intent is to delete the imported data at the end of the exercise to guarantee a design that has no features attached to the original data. Parametric body design principles are covered along with Pro/SCAN-TOOLS functionality to introduce the mixture of parametric surface modeling and free-form surface modeling. Tips and techniques are covered to highlight the pros and cons of both free-form and parametric surface modeling within a design project. Understanding the pros and cons will enable you to make engineering choices as to the trade-offs involved in using either approach.

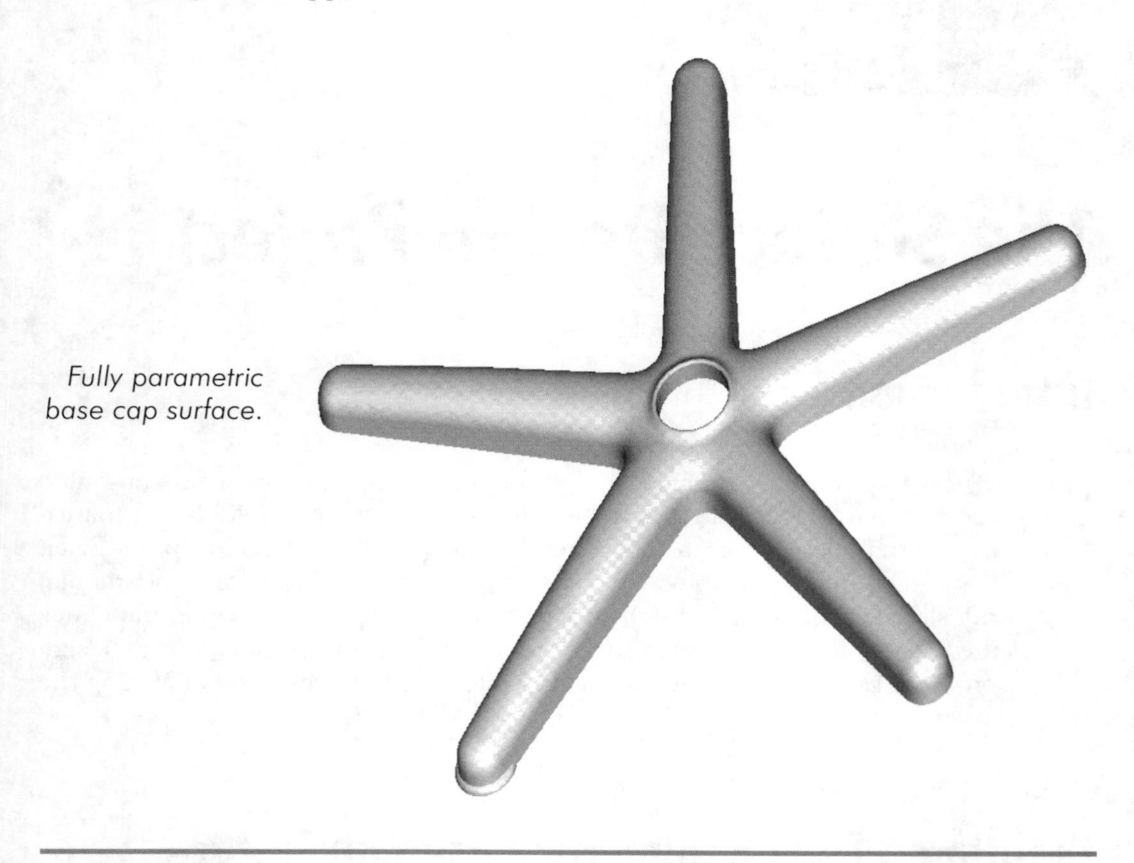

Fully parametric base cap surface.

Leg Development

Create default datum planes and then read in the imported data as datum curves from file or scan curves. Begin to think about how the original part was developed, or if you are working with scan data, try and visualize the overall patch structure of the part. Attempt to develop in your mind how certain patches were trimmed from overbuilt extended curves and surfaces. The starting slab surfaces are the most crucial toward lending flexibility for future changes.

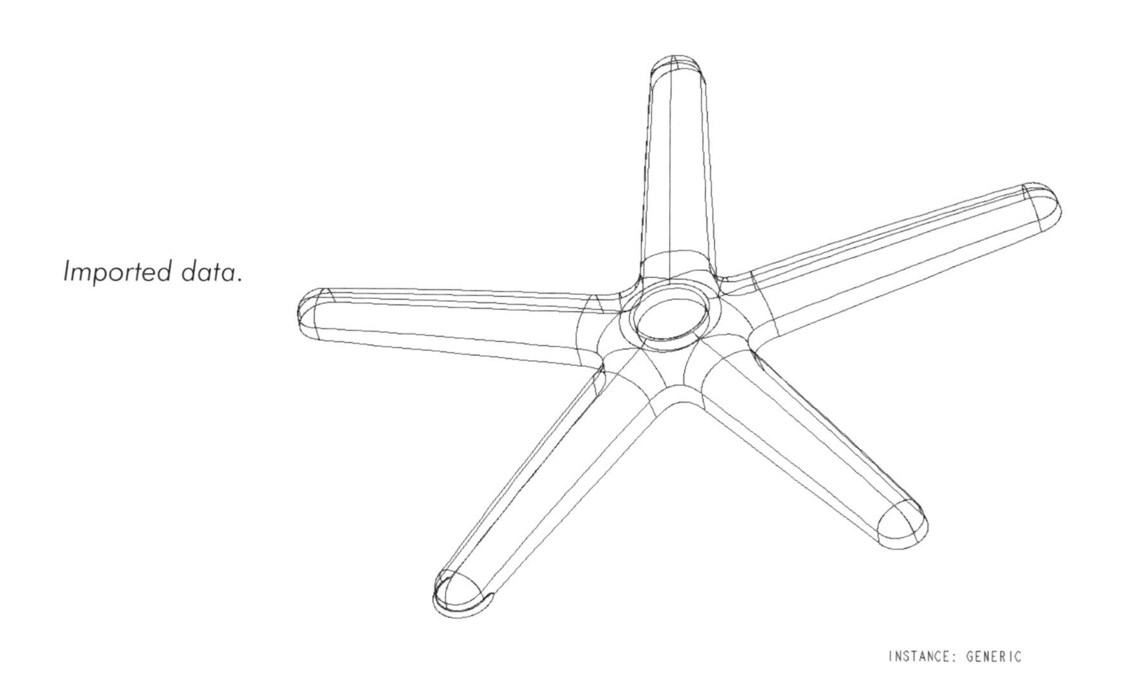

Imported data.

Isolate one arm of the base cap in order to simplify the amount of geometry on the screen. Begin to lay out the new parametric patch structure. When creating new datum curves, remember to tie all the slab curves to datum planes, points, and axes in a logical manner for maximum flexibility. Next, do not forget to rename datum planes and place them on corresponding layers.

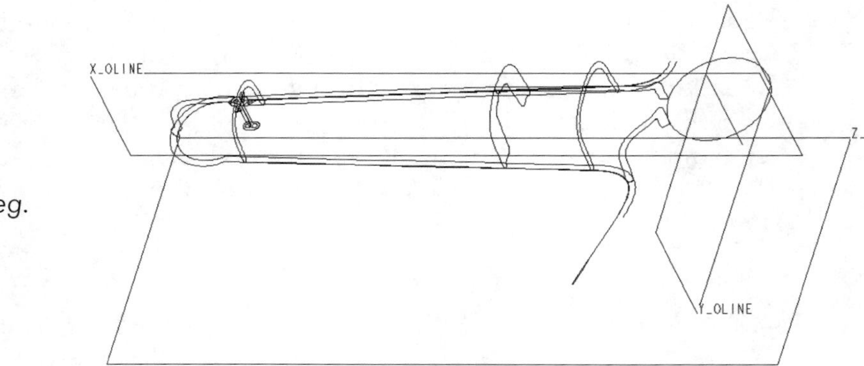

Isolated leg.

On the center plane of the arm, begin to lay out slab conics curves that represent the side view shape of the arm. This view of the arm consists of a top slab conic and a side slab conic. Check the smoothness and quality of the curve as described in Chapters 10 and 11. Layer away IGES data so that the data are displayed only when required.

When first placing the conic, try and create the curve in the closest possible position to the respective underlying data. Place the dimensional constraints and adjust accordingly to try and capture the shape of the part. Concentrate on making a smooth curve as close as possible to the reference data, but do not spend all day trying to duplicate the reference data because this would be very inefficient. Our intent is to simply capture the patch structure at this point and fine tune the smoothness and deviation later in the design.

✔ **NOTE:** *Use the Geom Tool Move option to dynamically move the end points to assist in shaping of the curve.*

Slab curve construction.

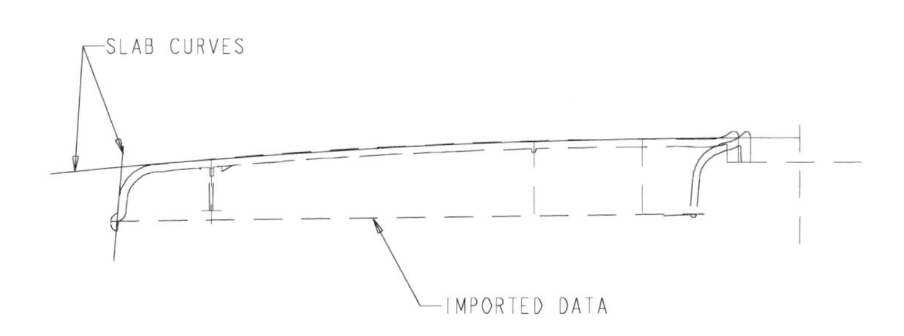

Create the nose curve that blends from the top slab curve to the side slab curve (conic or spline).

Use tangency constraints to connect the conic nose curve. Adjust the rho value or slopes to shape the position of the curve.

✔ **NOTE:** *In order to place or locate the nose curve, datum points and axes must first be created depicting the start and stop locations of the curve.*

Nose curve.

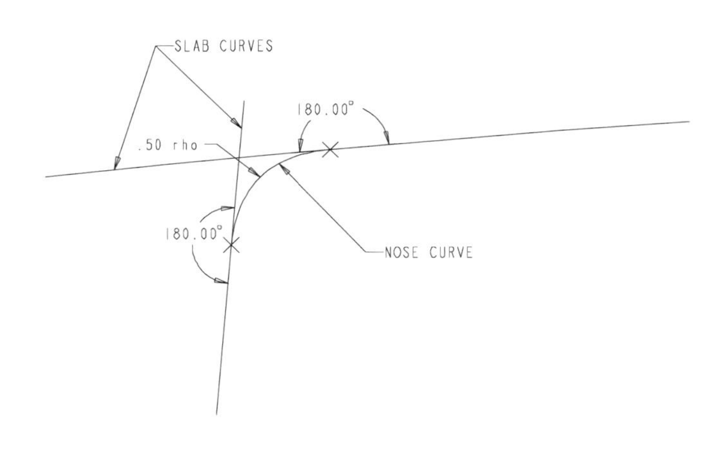

Create a tangent line curve parallel to the top curve. Make this curve an independent feature that passes through the vertex of the nose curve and the side slab curve.

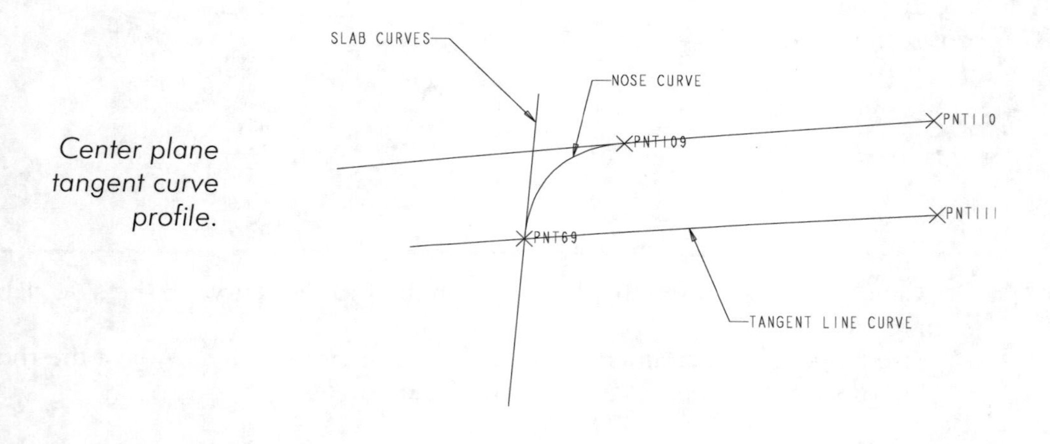

Center plane tangent curve profile.

Create a curve which represents the bottom of the cap. Next, create a datum plane with angular controls through the top nose vertex normal to the center plane. Rename the plane *nose_side_tang*.

Create the profile for the bottom of the cap in the plan view. Locate the start and end points of the conical curve by sketching center lines aligned through the end point of the side bottom profile curve. Adjust the curves accordingly for smoothness and deviation.

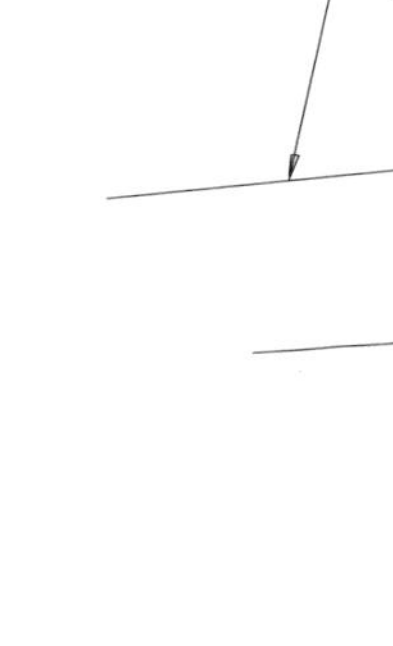

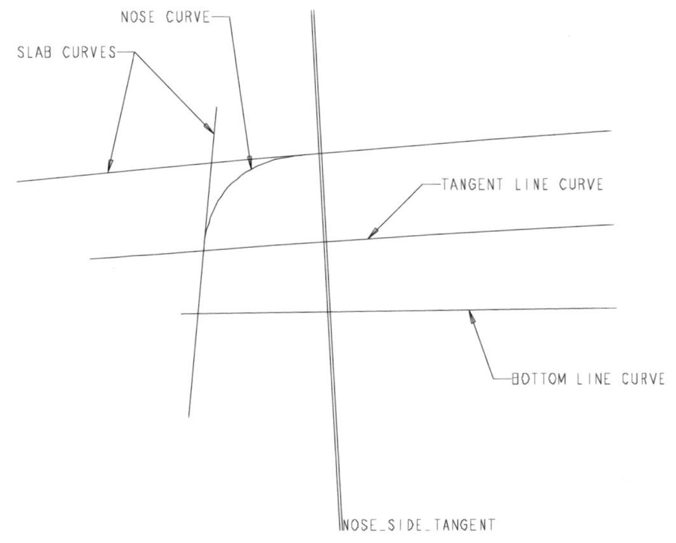

Center line bottom curve profile.

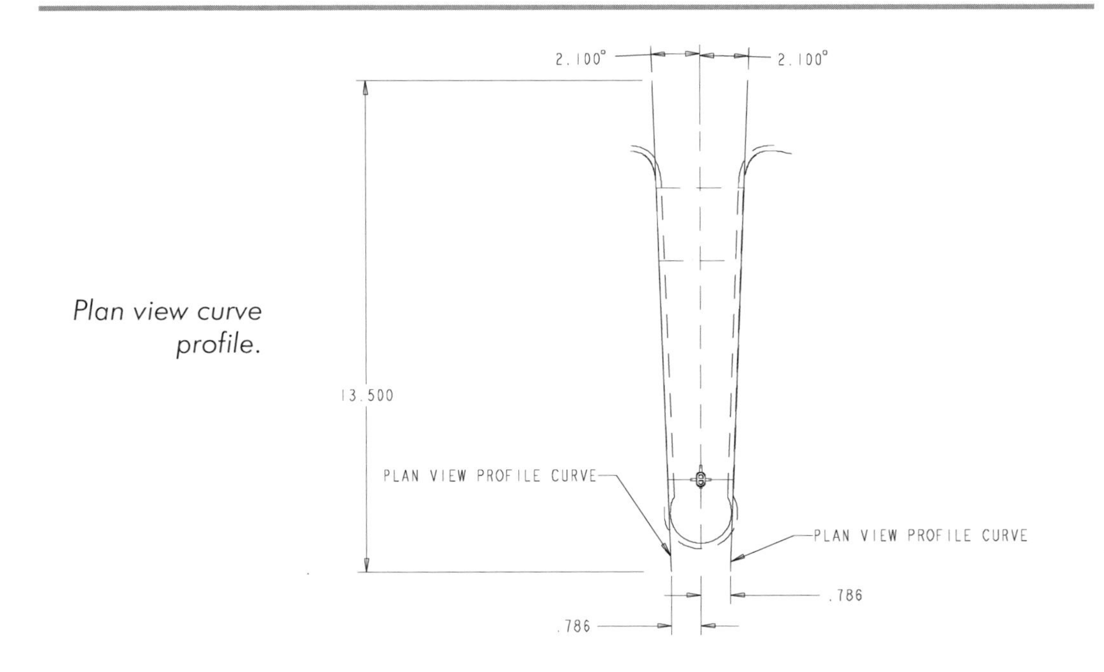

Plan view curve profile.

Create a 3D curve by combining the plan view profile with the side bottom profile curve. Use the datum curve 2 projections functionality.

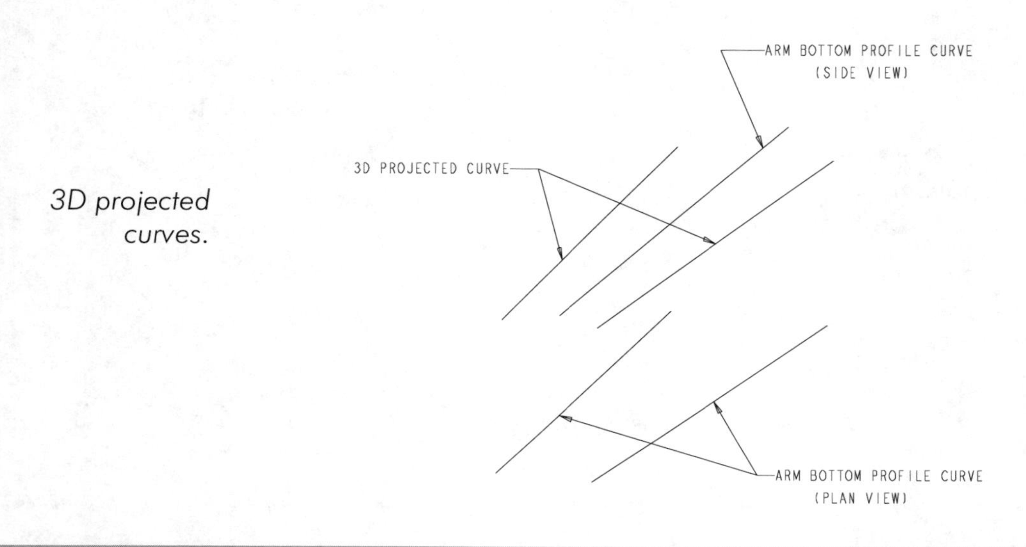

3D projected curves.

✔ **NOTE:** *At this point you should fine tune both the deviation and smoothness of the curve with respect to the reference data. Check the quality of the 3D curve because this curve is used for generating surface data.*

Sweep a conical slab surface along the 3D curves. These surfaces represent the side surfaces of the leg of the base cap.

Measure the smoothness and deviation of the surface and adjust accordingly. Sacrifice the deviation (within reason) to the scan in favor of a smooth surface.

Project (drop) the planar side view tangent line curves on the newly created side slab surfaces. Create a datum curve by intersecting the side slab surfaces with the datum plane named *nose_side_tang_cont.* Create datum points and datum axes (both directions) at the intersections of the vertical section curve, projected side tangent curves, and 3D curves.

These axes will help in locating the curve sections that define the lower half of the nose.

Swept slab surface.

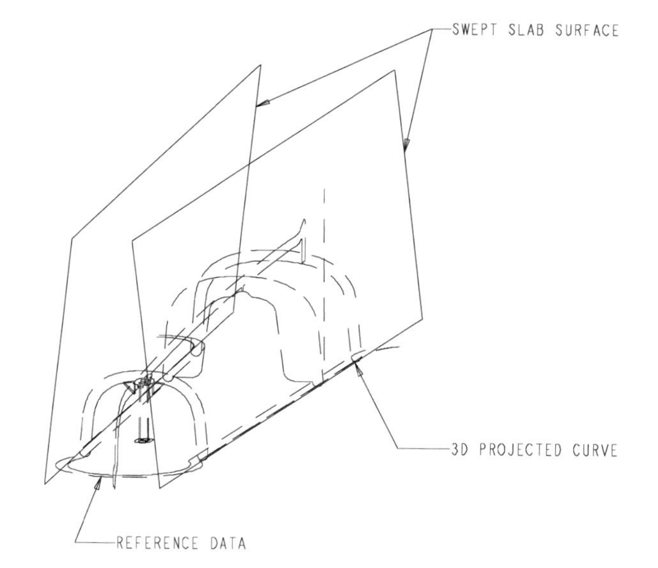

Projected side tangent curve on swept slab surface.

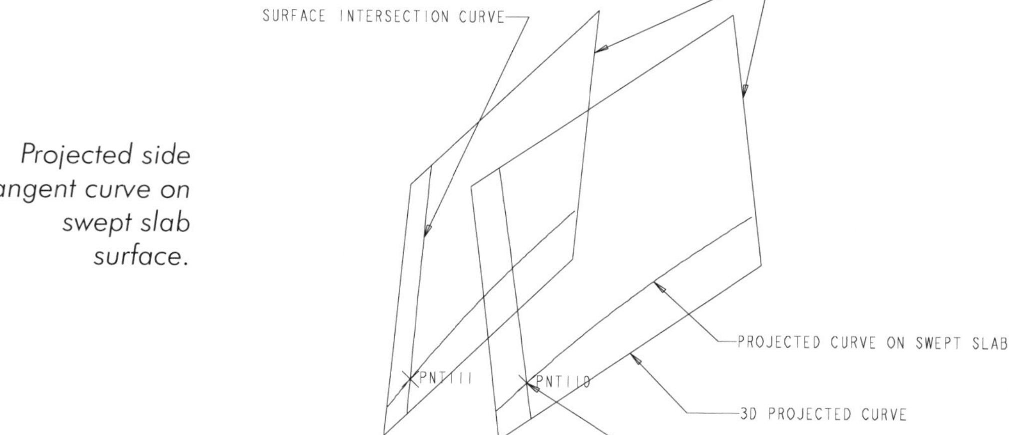

Create conic curves in the other direction from the center profile curve to the side intersection curves. This action closes the loop for creating the top surface of the arm.

✔ **NOTE:** *Sketch the first curve of the plane nose_side_tang_cont which passes through the nose top vertices and the side intersection curve. Hold tangency constraints to the predefined datum axes. Sketch the other curve through the end points of the 3D curves orthogonal to the master datum planes.*

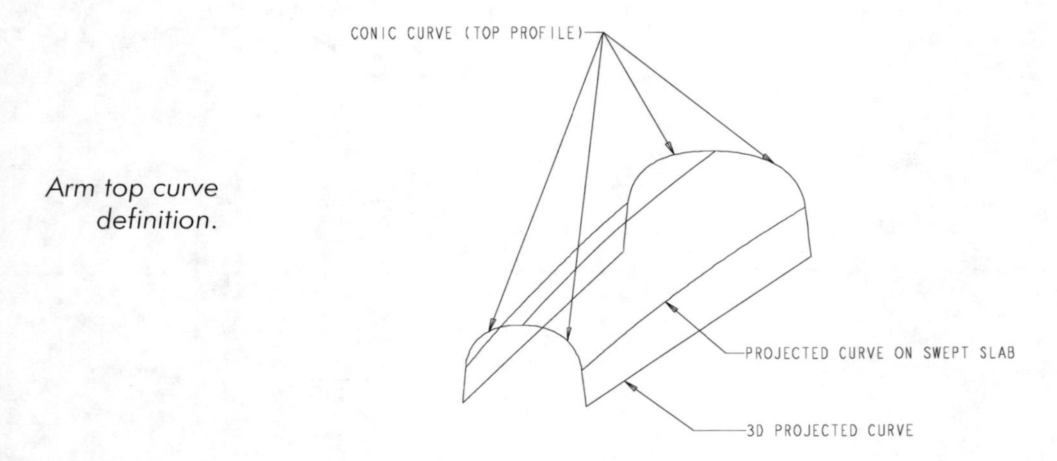

Arm top curve definition.

Develop the lower portion of the nose curve patches by sketching conics on the corresponding tangent planes. The sketching planes are defined by creating them on the fly through the axes at the end of the curves and the vertices on the center line nose section.

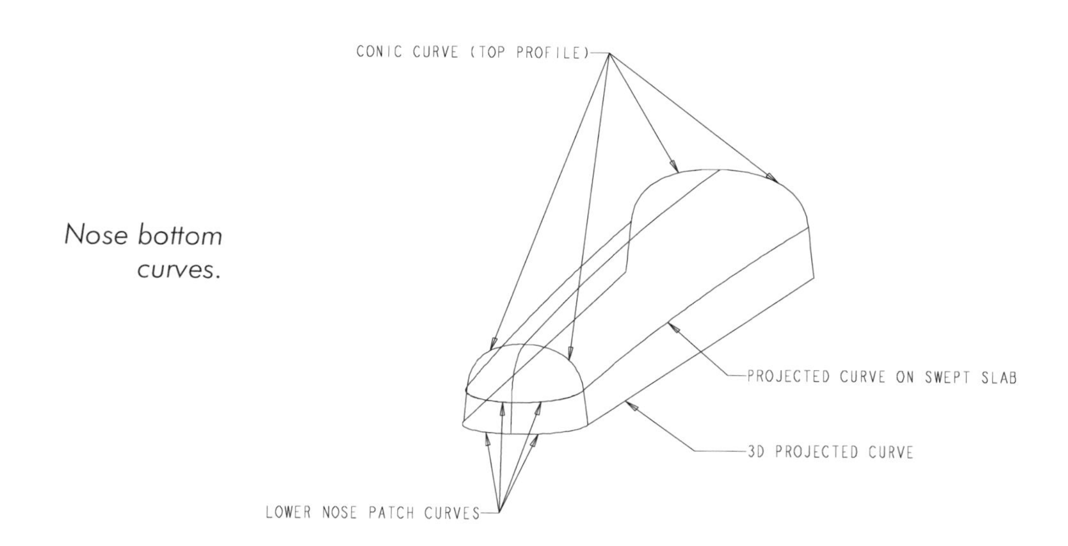

Nose bottom curves.

Surface the right top half of the arm followed by the right lower patch located below the nose. Hold tangency conditions to the swept side surfaces and dummy extruded surfaces of the master center line sections. Layer away reference extruded surfaces. Surface the corresponding left side while holding tangency to the side sweep surface and the newly created right-side surfaces.

A clean top surface is important because we will be holding tangency constraints on the nose surface. Because the nose surface is a three-patch surface, it is important that the direction of the curves selected is correct for a good blend.

Surface the triangular nose patch holding tangency to the corresponding edges.

At this point, you need to select the top and bottom curve first as one direction, with the single center section curve being the last. Hold tangency constraints to the top and vertical surfaces. Hold the center section normal to sketch or create a dummy surface that is extruded from the curve normal to the center plane. This will allow tangency constraint to the extruded surface. Layer away extruded reference surface. Finish the other nose surface and add tangency constraints to all sides.

Use the Surface Transform Rotate option to rotate the quilted surfaces five times every 72 degrees. Any changes made to the original leg will propagate across the entire part. Quilt the five legs together to form a master leg quilt.

Boundary surfaces.

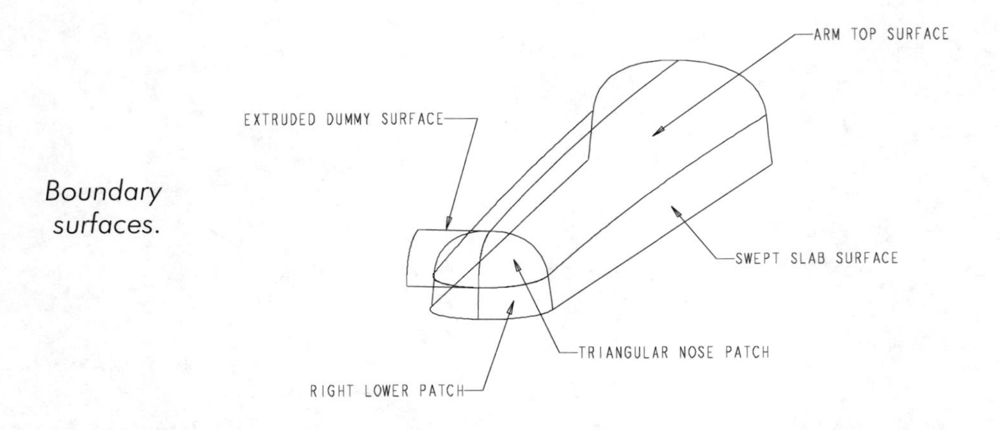

The following illustration shows the final results.

Rotated surfaces.

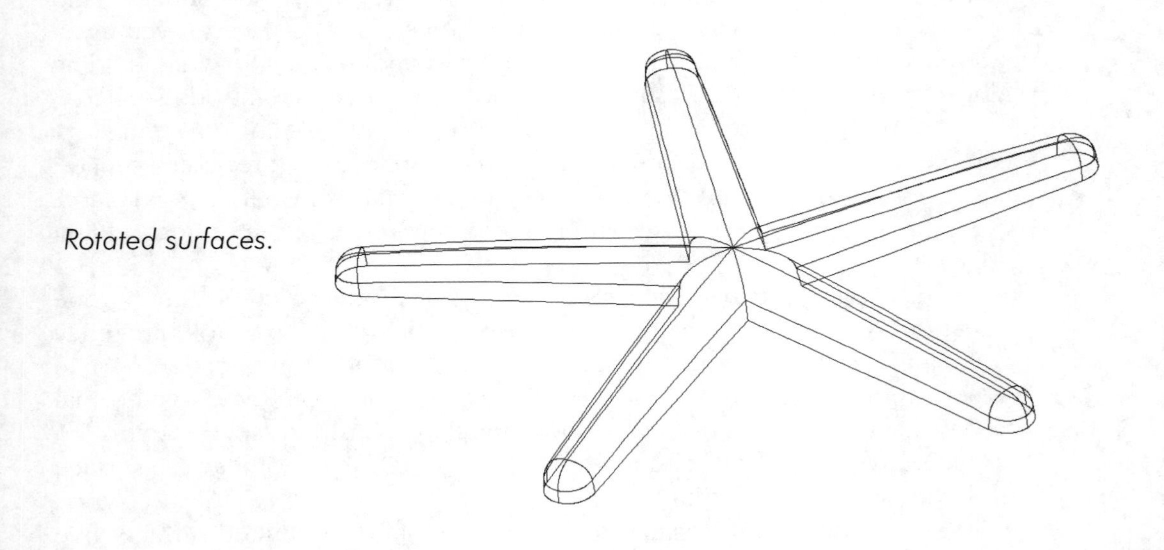

Web Development

Create datum points along the center line section of two legs of the base cap. Make sure that the points are equidistant from the center line of the part. In the plan view, create a datum axis that passes through each point normal to the plan view (Z_Plane). Create datum planes through the datum axis with an adjustable angular constraint. Verify that the planes are normal to the Z_Oline plane.

Create a datum curve on each leg by cutting each plane through its respective leg. These curves represent the tangent lines of the blended web surface.

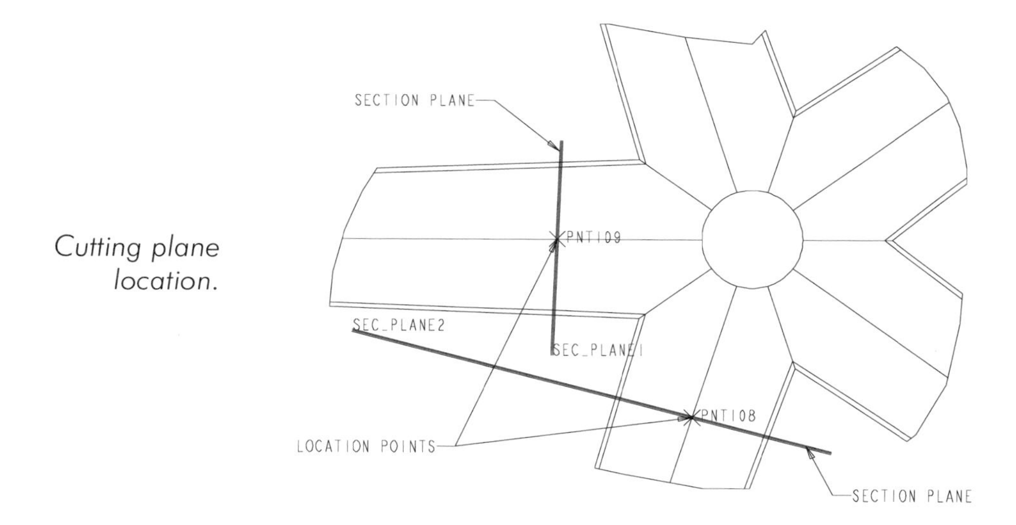

Cutting plane location.

Develop top and bottom blending curves (conics) located between the base of the two legs that represent the lower surface patch. Make sure to hold tangency constraints on the connecting blended curves. Graphically and analytically check the curvature of each curve at the point where it blends into the lower leg. Tangency conditions should be held without C2 continuity conditions.

Verify that datum points and axes are established prior to creating the new curves. Develop the top curve by creating a datum plane on the fly through the normal axis vectors located on the end points of the top curve.

When the project is complete, insert the mode prior to the tangent blended curves. Through the use of Pro/SCAN-TOOLS, add C2 continuous style curves that represent the blended curves. Reroute the web patches to the style curves and check the quality of the blend. The web surface will have a higher rate of continuity blending across the common seam. The only trade-off is that the newly created style curves are not parametric. The basic rule is to create a parametric curve that updates throughout multiple design iterations and then replace with C2 continuous curves before cutting tools.

Surface from boundaries using tangency conditions.

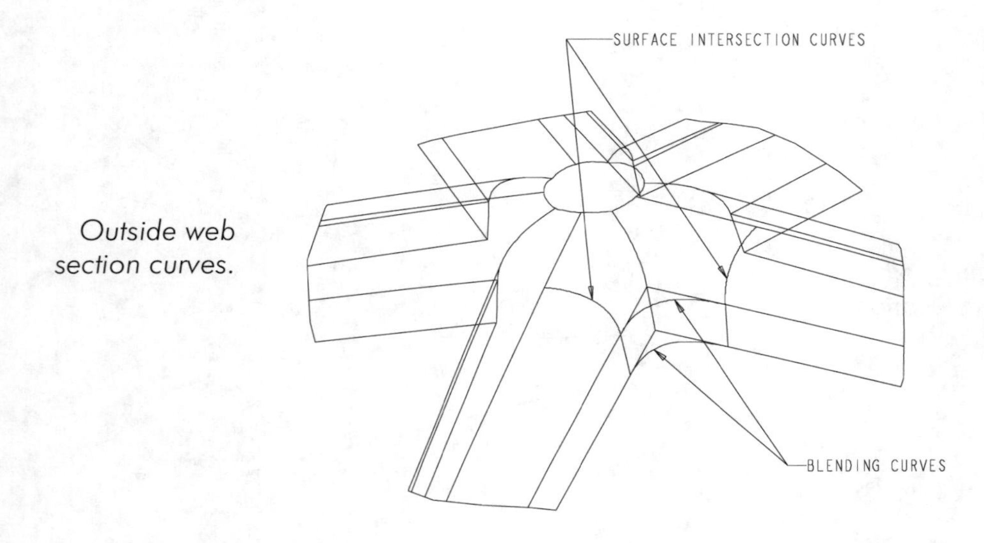

Outside web section curves.

In the plan view, create datum planes that will be used to add intermediate control curves. These planes are located through controllable datum points located on the top boundary edge of the lower web patch surface. Verify that these planes have angular constraints and are normal to the Z_Oline plane.

Create datum curves by intersecting the new planes with the main quilt and the lower web patch surface.

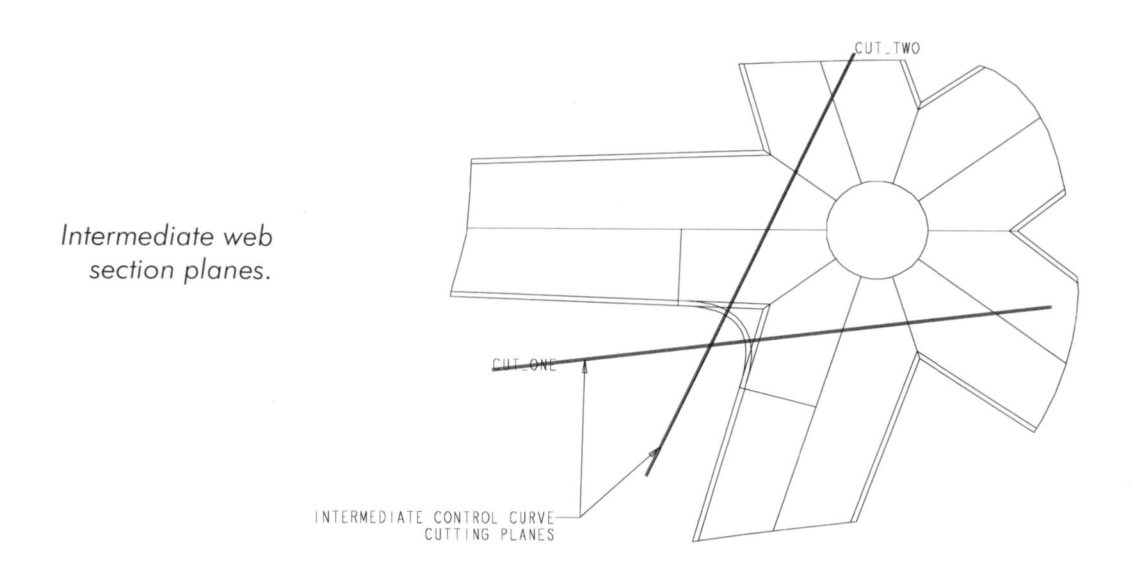

Intermediate web section planes.

Create connecting curves between the curves generated from the intersecting surfaces. This breaks the upper web surface into three individual patches. Hold tangency constraints on these curves to previously defined datum points and axes.

Create the connecting top curve of the upper middle web patch. This curve is generated by creating a datum plane on the fly through the normal vector datum axes located at the ends of each upper web patch.

Surface the outer upper web patches first while holding tangency conditions to each adjacent edge. Surface the upper middle patch last and hold tangency to the side and lower boundaries. Create one master web quilt by merging the four independent patches together.

Transform the master web quilt five times every 72 degrees and merge with the master leg quilt. Check the quality of the surfaces and the blended intersections through graphical and analytical checks (Porcupine, Max Dihedral, Rendering, and Cutting Sections).

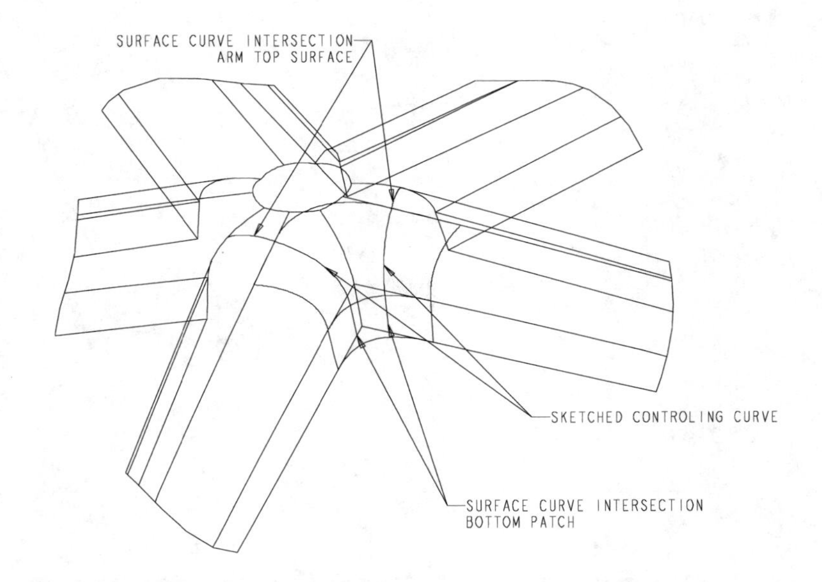

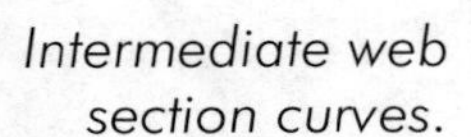

Intermediate web section curves.

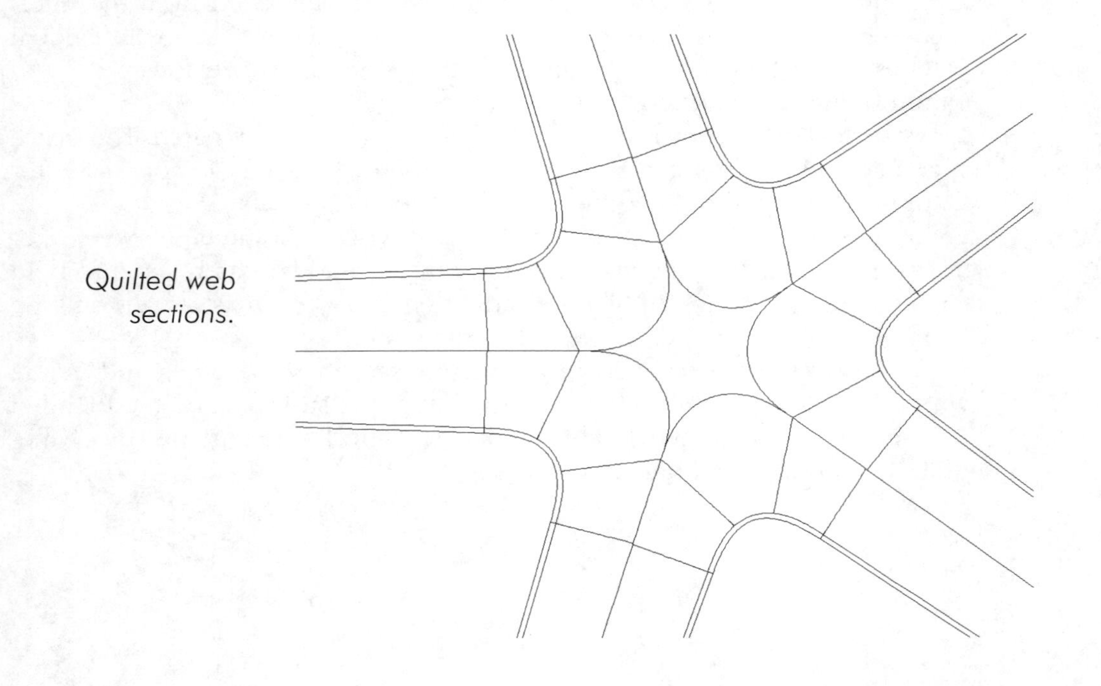

Quilted web sections.

Ring Development

Layer away the surfaces and begin to focus around the center ring area.
Create a revolved surface that represents the cross section of the inner hub of the base cap. Use the imported data to guide the shape of the cross section.

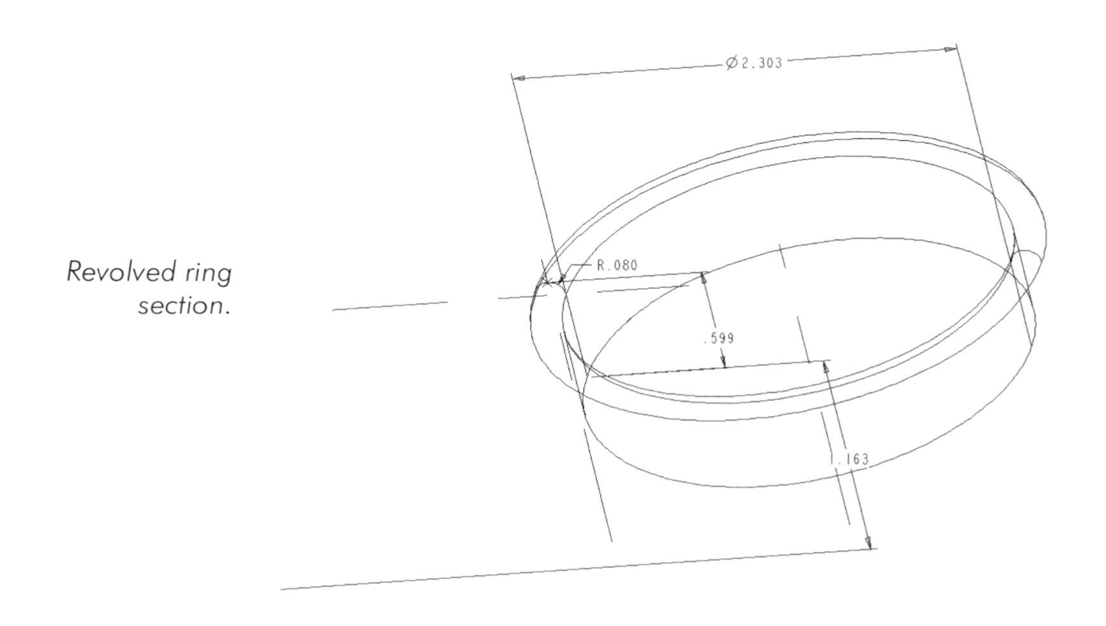

Revolved ring section.

In the plan view project a curve (circle) on the fillet surface patch. This will be the tangent line for the blended surface that belongs between the revolved patch and the main surface quilt. Next, in the plan view, project another curve (circle) on the main quilt that represents the tangent line for the blended surface as it washes out into the main quilt.

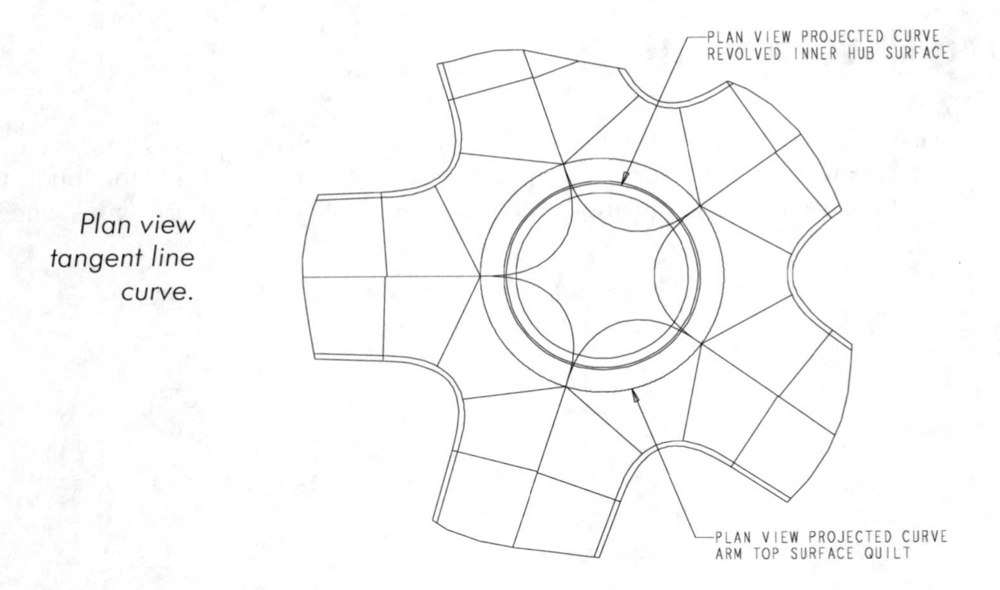

Plan view tangent line curve.

Cut sections through the revolved surface at the center of each leg. These sections should intersect the curve that was projected on the revolved surface. Create datum curves from the top edge of the quilt of the revolved legs. These curves will be used as start and end points for the blended fillet surface.

Begin to construct the fillet curve between the revolved surface and the main quilt. Verify that you have points and surfaces established prior to creating the blended curves. These axes are vital for holding tangency to the sketch conic or spline curve.

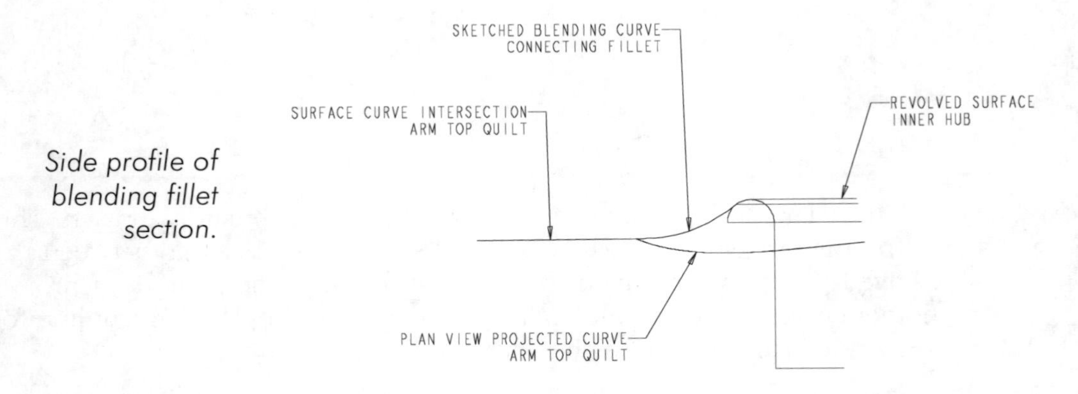

Side profile of blending fillet section.

Surface the fillet curves independently and then quilt together to form a master blended quilt between the revolved surface and the main surface quilt. Merge the fillet quilt with either the revolved surface or the main quilt. Complete the quilt with the other surface.

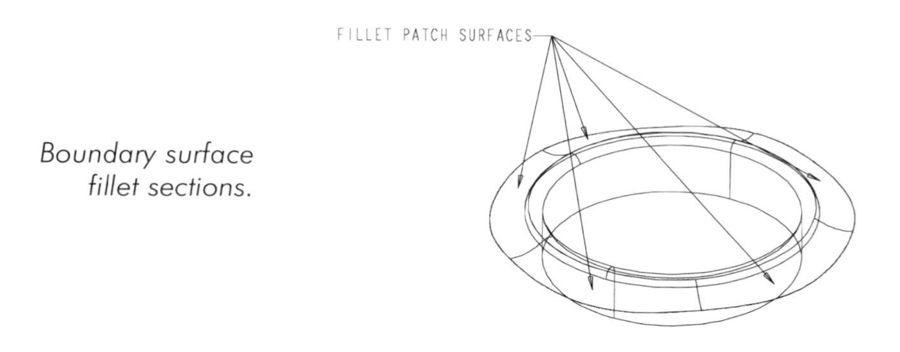

Boundary surface fillet sections.

Completed quilted section.

Check the overall quality of the blended surfaces as described earlier and in Chapters 10 and 11. Parametrically change variables to compromise surface deviations and surface quality. Create a solid thin feature and finish the final details of the model.

Summary

The example in this chapter demonstrates how to break down a design into a logical patch structure. The patch structure is simple and offers flexibility in multiple design changes in that the model is fully parametric and may be flexed for future changes. In summary, our basic philosophy is to keep it simple and tie the wireframe together in a manageable way though the use of datum points, axes, and planes. These techniques may be applied for surfacing to scan or IGES data, or developing concepts from scratch to an artist's renderings. Keep in mind that surfaces are as good as the underlying curve geometry.

PART IV

Four

Assembly Functions

Following A Blueprint Into Assembly Mode

Introduction

As in developing parts, you must consider many aspects of the entire product development when developing assemblies. The most important aspect is creating and following the design blueprint (design intent). When creating assemblies, your main focus should be on breaking down the function and package requirements for the design. The objective is to develop assemblies which contain logical setups to fulfill concurrent applications including designing in assembly, drawings, and manufacturing. Design changes are inescapable; therefore, assemblies should be flexible (susceptible to easy change or modification). In addition, you should take advantage of Pro/ENGINEER tools for managing large assemblies. In Chapters 13 through 16, we investigate examples illustrating development considerations and techniques to create assemblies.

Concurrent Engineering for Setup and Assembly

Setting up the assembly while considering the requirements for the product's function and package can help you achieve *concurrent engineering*. Concurrent engineering is another term for product development when all disciplines work together in a virtually simultaneous manner. Pro/ENGINEER's single database structure enriches concurrent engineering, especially at the assembly level. The assembly becomes an extension of the design process, as it should.

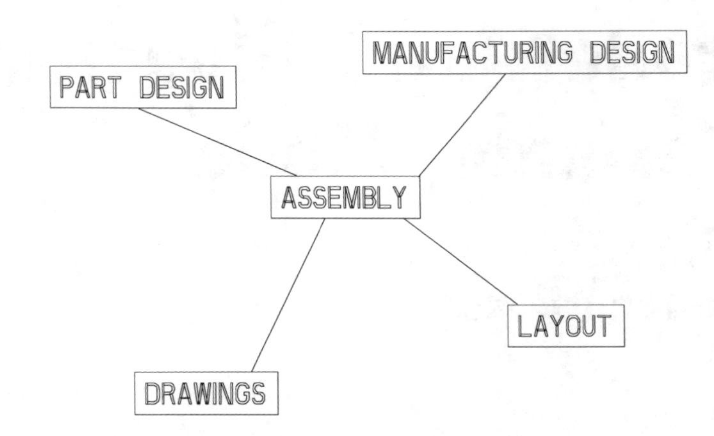

Many disciplines working together concurrently.

The blueprint serves as the hierarchy structure or bill of materials for the assembly. The blueprint for an assembly should reflect the explicit procedures needed to put parts and sub-assemblies together. Next, the blueprint can help you formulate relationships between parts or sub-assemblies. Therefore, as you make changes to the assembly all other disciplines react accordingly, even if the assembly changes are made at different levels of the hierarchy (i.e., subassembly or part level). Advanced users will often create this blueprint by sketching out the basic structure or by mentally formulating the structure.

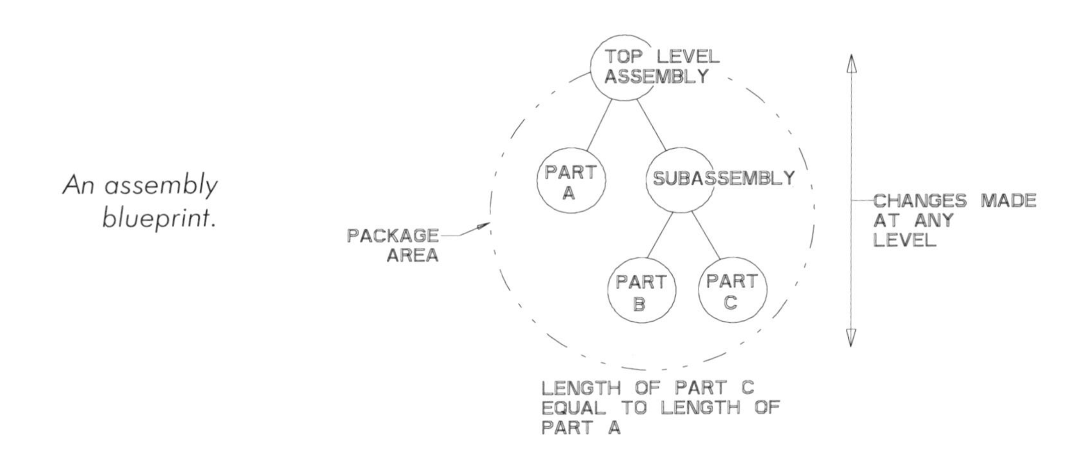

An assembly blueprint.

As you follow your blueprint into assembly mode, the following procedures will help you accomplish concurrent engineering.

- ❑ Whenever possible, assembly component placement, and parent/child relationships, such as mating or aligning parts, should be reviewed by all other disciplines involved.
- ❑ When designing a new part in an assembly, advanced users most often elect to first create the new part in Part mode with default datums. The new part is then assembled using the default datums. Consequently, you can explictly work on the new part in the assembly, thereby minimizing the chance for developing unwanted references and establishing a useful starting point for the new part being designed in the assembly.
- ❑ When setting up assembly features or reference geometry (Datum Planes, Datum Points, Assembly Cuts, etc.), you should keep in mind whether the assembly will need a drawing. Some focus on the placement and design of the assembly features can aid in the creation of the drawing. This simple tip can save an enormous amount of time when creating the assembly drawing.
- ❑ When creating an assembly feature, have a specific goal in mind. Think ahead about how you wish your assembly features to act and react within the assembly. Many features created in Assembly mode refer-

ence other levels, such as individual parts or other assemblies. These references should be considered and reviewed with the other involved disciplines.

- When creating dependencies or relationships between parts or subassemblies, care should be taken to ensure that relationships are completed in the order they are solved. The sequence for solving assembly regeneration follows:

  (1) Modified parts are regenerated first in the order which they were assembled.
  (2) In a modified part, feature relations are solved in the order in which they were created. This may include Evaluate features.
  (3) In the modified part, the Sketcher relations are then solved.
  (4) After each modified part has regenerated, Assembly feature relations are solved in the order in which they were created. This may include assembly Evaluate features. Assembly Sketcher relations are then solved.
  (5) Finally, the assembly components are solved and placed.

- Understanding whether a particular feature is important or necessary for manufacturing is also very important during the development of an assembly. Two examples of thinking ahead for manufacturing appear below.

  (1) Adding assembly cuts to represent machining operations. This operation may be performed in a manufacturing environment after some assembly has been completed, but not at the part level.
  (2) Adding assembly cuts to represent drilling operations. This operation may also be performed in a manufacturing environment after some assembly has been completed, but not at the part level.

✔ **NOTE:** *As of Version 15.0 of Pro/ENGINEER, assembly features which may affect the part, such as cuts and holes, are stored with the assembly rather than with the part. This enhancement eliminates the possibility of an external part reference to an assembly.*

The above examples of thinking ahead for manufacturing point out the importance of thinking about other disciplines. Such thinking ahead will help to streamline product development. In addition, you will be getting what you need and expect during development concurrently with other disciplines, thereby avoiding costly surprises and changes. This practice is beneficial only if the blueprint or development intent is followed.

Flexing the Assembly

One of the most enlightening tips that any user could receive is to flex the assembly during development to ensure assembly quality. Flexing the assembly is simply the act of making modifications and changes to the assembly during development. A quality assembly fulfills all other discipline requirements. In addition, a quality assembly can be easily changed through Redefine or Modify. During the flexing of the assembly, problems or failures may occur. Assembly failures are generally welcome because they help identify problems that may not have been considered during blueprint development. The failures give you an opportunity to fix the design up front during development rather than later when revisions may be more difficult and costly. Basic guidelines for flexing an assembly appear below.

- Redefine or Modify to determine whether the assembly will act or react according to the design blueprint (intent).
- Play the "what if" game. As you add parts or sub-assemblies to your assembly consider different function or package scenarios that may occur.
- Redefine or Modify different values according to the different function or package scenarios and then Regenerate.
- Make flexing modifications after every five to ten features are added to the assembly, and Regenerate. Recall that each component added to the assembly is an assembly feature.

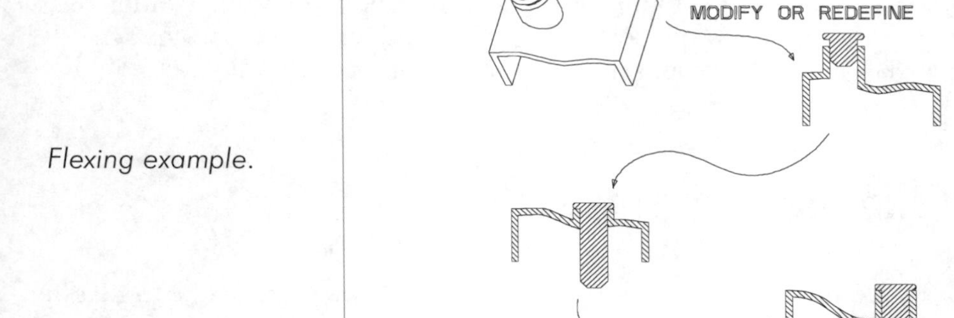

Flexing example.

Managing large assemblies can be difficult when you are not familiar with certain tools available in Assembly mode. When you take advantage of Simplified Representations and Interchange mode, you will find that large assemblies are much more efficient, and thereby lead to more productive development. We will examine some of these methods in subsequent chapters.

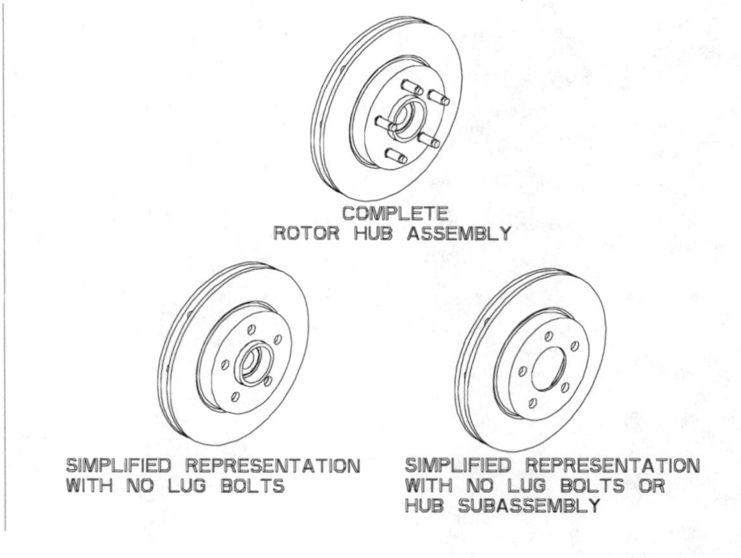

Managing large assemblies using Simplified Representation.

Summary

During assembly development, the most important item is the product's design blueprint or intent. When a focus is set up, established, and maintained you can efficiently develop elegant assemblies, both simple and complex. Other applications can be run concurrently and costly surprises or changes during development will be limited. Flexing an assembly will help ensure design intent and point out possible problems in the function and package of the assembly. Combining focus and flexing will result in a quality assembly that fulfills your design blueprint and is easily modified or changed.

Kinematics: Making Assemblies Move

Introduction

When developing assemblies, a common requirement is that various components of the assembly move in relation to one another. The movement of the assembly can be described as *kinematics*. In this chapter we will illustrate two examples of creating assemblies that move. The examples will explain various techniques to create kinematics. The first example, which reviews basic requirements in developing assembly movement, will describe a two-dimensional linkage similar to a locomotive. The second example will focus on developing complex three-dimensional movement using an automotive suspension as the blueprint. This example will also show how to automate the kinematics through Pro/PROGRAM.

Approaching the Moving Assemblies Design Problem

Throughout the book we have described the importance of developing a blueprint (design intent) when approaching a design problem. The first example for kinematics will describe the "best" technique for developing kinematics assemblies. The technique establishes the blueprint for a kinematics design.

When reviewing the needs for the kinematics assembly, you must first understand how the parts in the assembly will work together. Our first example of a linkage system contains three parts. The function of the kinematic assembly is driven by a rotating wheel. As the wheel rotates, a fixed-length arm rotates, with one end attached to the drive wheel and the other end attached to a horizontal slide.

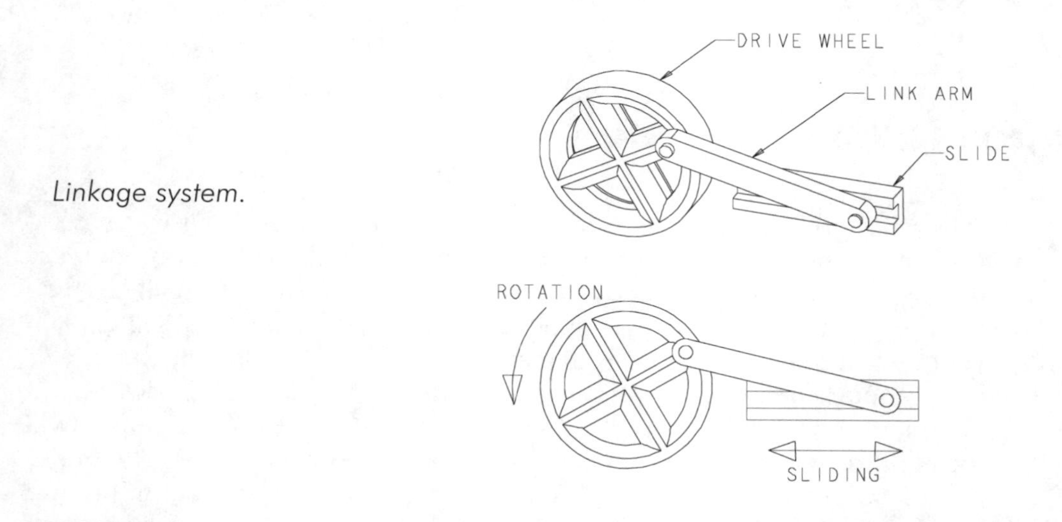

Linkage system.

The three components—drive wheel, link arm, and slide—will be assembled using an additional part called the "skeleton." The skeleton part will be the base from which the kinematics assembly is controlled. The use of an additional part is a unique approach called the "skeleton technique."

How the Skeleton Technique Works

The skeleton technique involves the use of Datum features and Surfaces to develop all types of kinematic assemblies. To illustrate this technique we will set up the skeleton part for controlling the linkage system.

Begin by creating a new part with default datum planes called *skeleton*. The first two features in the linkage example are extremely important because they establish how the kinematics assembly will be driven. The initial feature is a Datum Curve sketched as a circle. The circle defines the diameter of the drive wheel.

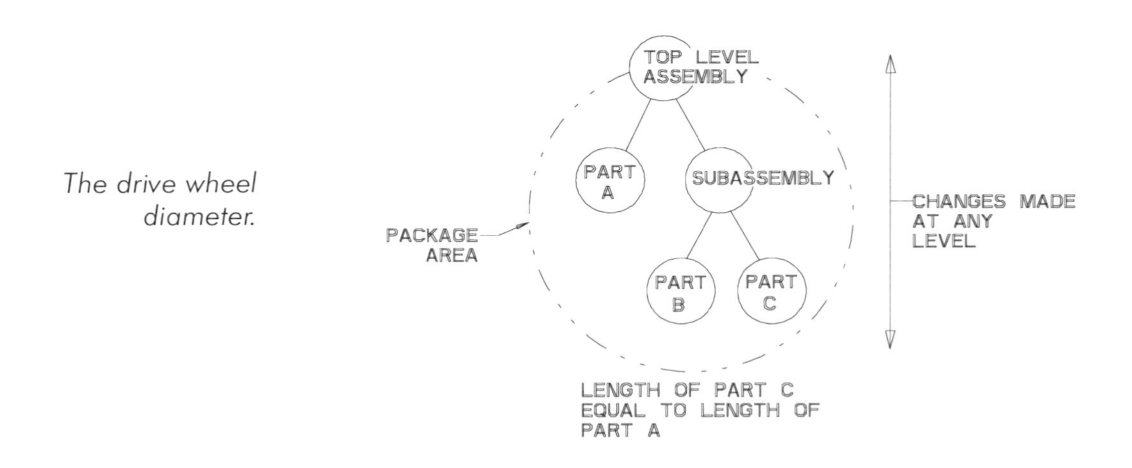

The drive wheel diameter.

The second feature is also a Datum Curve. This feature will contain one dimension (angular), which will rotate the kinematics assembly. The feature is sketched using the previous sketching plane, the same plane as the circle. It is defined with one angle and aligned to the center and outer diameter of the circle. The completion of this feature finalizes the driving features for the linkage.

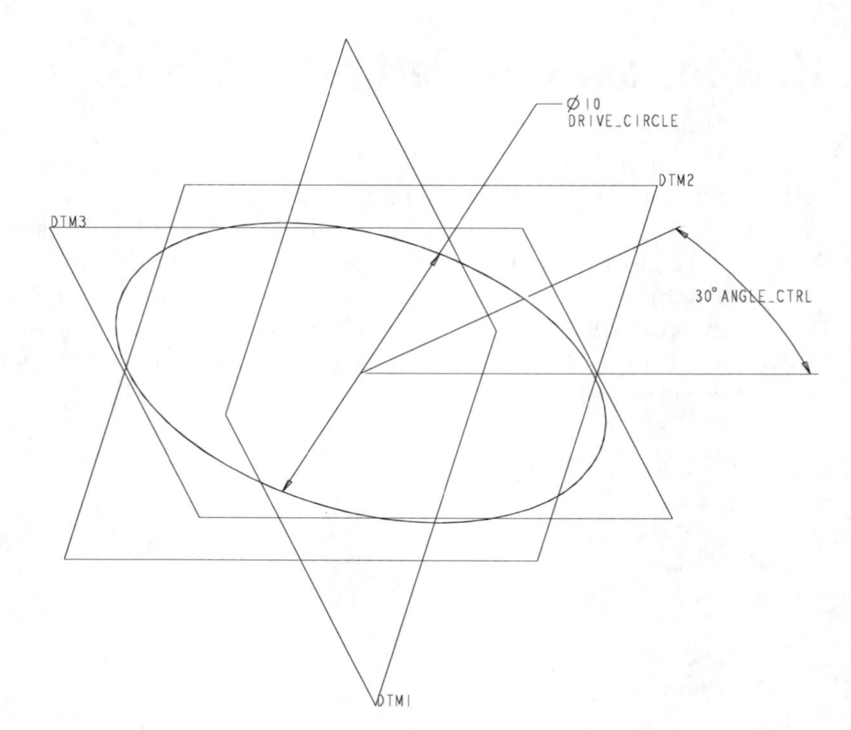

The angular drive dimension at different positions.

Only two additional features are required to complete the skeleton part. As with the driving curves, the next two features will also be constructed as curves. The slide feature is constructed next as a horizontal curve slightly below the center of the circle. The distance below the center of the circle has been determined through reviews involving all parties concerned with assembly design.

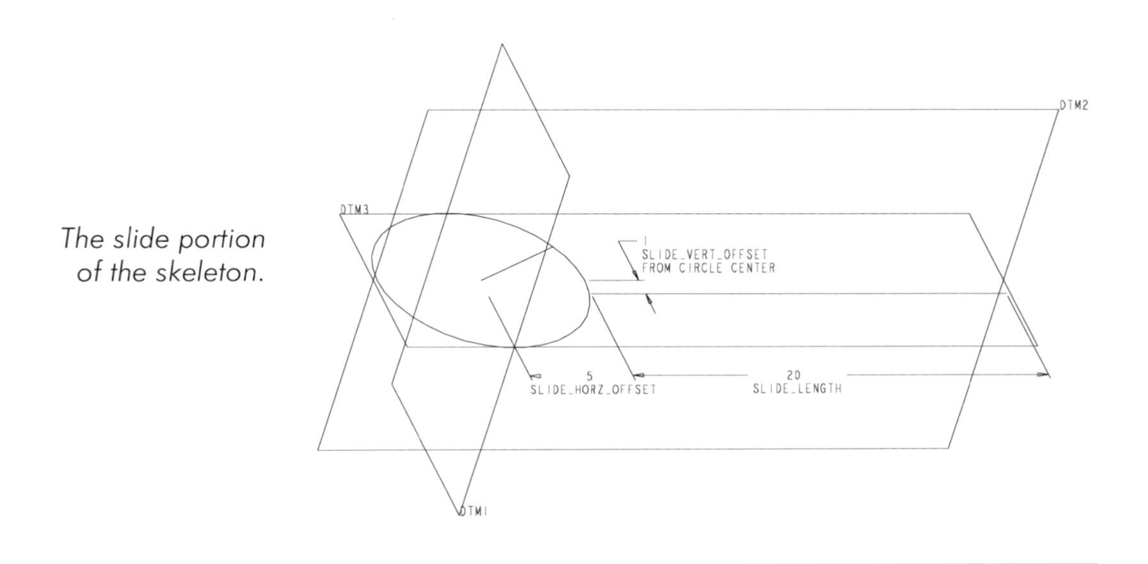

The slide portion of the skeleton.

The final curve will be sketched similar to the previous curves. This curve defines the fixed-length link arm. The link arm is sketched by aligning one end to the circle and angle curves, and aligning the other end to the slide curve.

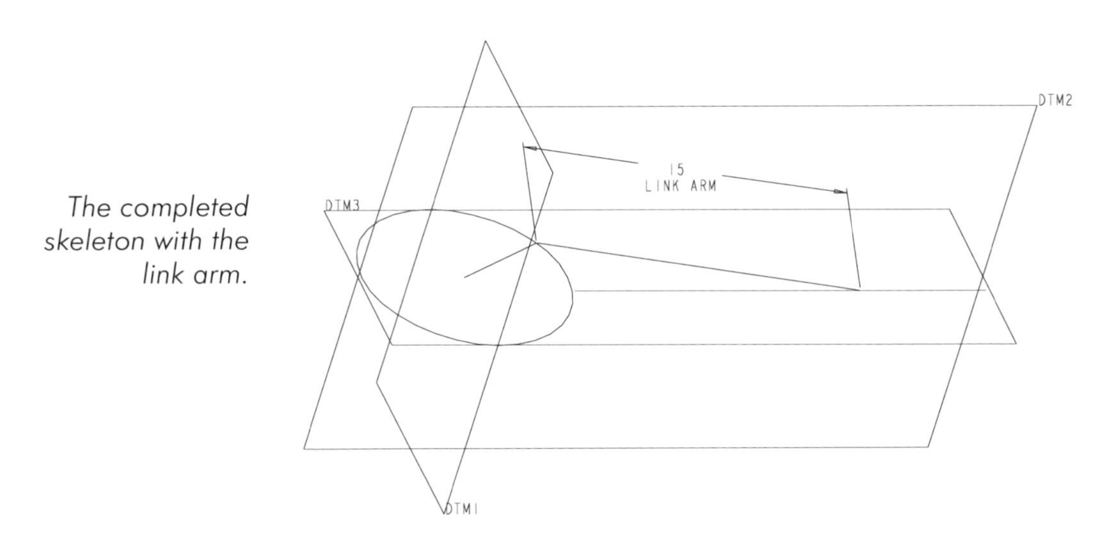

The completed skeleton with the link arm.

Creating the Kinematic Assembly

To create the kinematic assembly, we begin by placing the skeleton part into the assembly first. The order in which the rest of the parts are assembled is not important. However, advanced users tend to develop an assembly practice, such as assembling the parts as they relate to the development of the skeleton. Assemble the drive wheel, the slide, and the link arm.

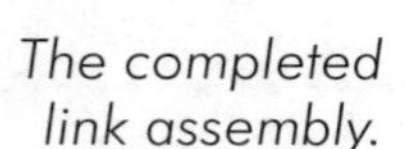
The completed link assembly.

Creating the Kinematics in the Assembly

The most efficient way to generate the kinematics in the assembly is to directly modify the skeleton part. Select **Modify → Modify Part → Skeleton** and modify the feature that controls the angle (*feature #6 - angle_ctrl*).

✘ ***TIP:*** *Take advantage of naming features at the part level. If the feature is named, you can select it directly from the menu by using* **Sel By Menu → Name**. *Important features of the skeleton such as the driving or controlling should be named for easy access. To name a feature* *in part mode select* **Set Up → Name → Feature** *and choose the feature you wish to name.*

Next, Regenerate the skeleton part from the Modify Part menu. Do not execute an Automatic Regeneration because a part regeneration will render the best performance. After regeneration of the skeleton part is complete, the parts reassemble to their new location. Efficiency will be quite evident as the parts quickly move to their next location.

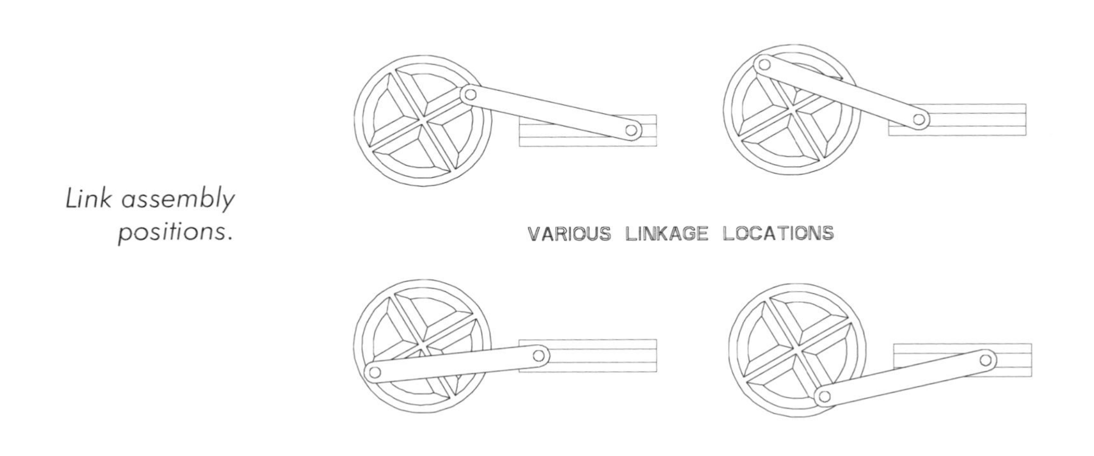

Link assembly positions.

Benefits of the Skeleton Part

Using the skeleton part yields many benefits. With large assemblies, you can use Simplified Representations and Interchange Mode to develop many different studies of the assembly movement. Assembly parts can be disassembled, suppressed, or blanked, and the linkage information is still available. Relationships can be developed between the skeleton and assembled part to create an even more powerful assembly.

Expanding the Skeleton Part into 3D

The first example used the skeleton part to describe the kinematics for the assembly in two-dimensional space. However, the functionality of this technique is not limited to two-dimensional space. Another example will illustrate how this technique can be used in three-dimensional space. For this example an automotive suspension will be described and then automated.

A Kinematic Short Long Arm (SLA) Automotive Suspension Assembly

An SLA suspension consists of three major components or subassemblies. The components of an SLA suspension typically lie on different planes. In fact, the skeleton part for an SLA suspension requires the use of Datum Curves, Datum Points, and Surfaces in order to function. This three-dimensional example will focus on the front left suspension of an automobile. The SLA skeleton expands upon the first example illustrating how joints and positions in three-dimensional space can be developed.

The blueprint to describe the SLA skeleton's function can be established via a few steps. The steps will illustrate how each feature is created and controlled. The illustrations will also describe the placement of the components within the suspension.

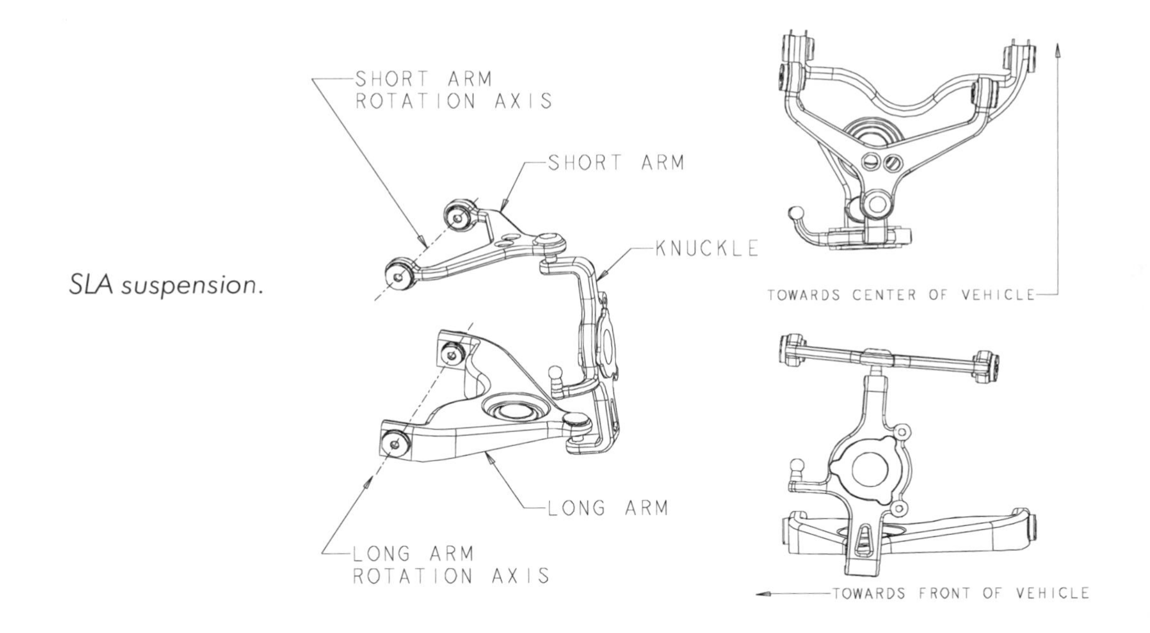

SLA suspension.

Initial Skeleton Design Position

The SLA suspension's initial location should be designed relative to the front center of the vehicle (commonly referred to as 0,0,0, or vehicle position). Coordinate Datum Points are used to accomplish this requirement, which is driven by engineering.

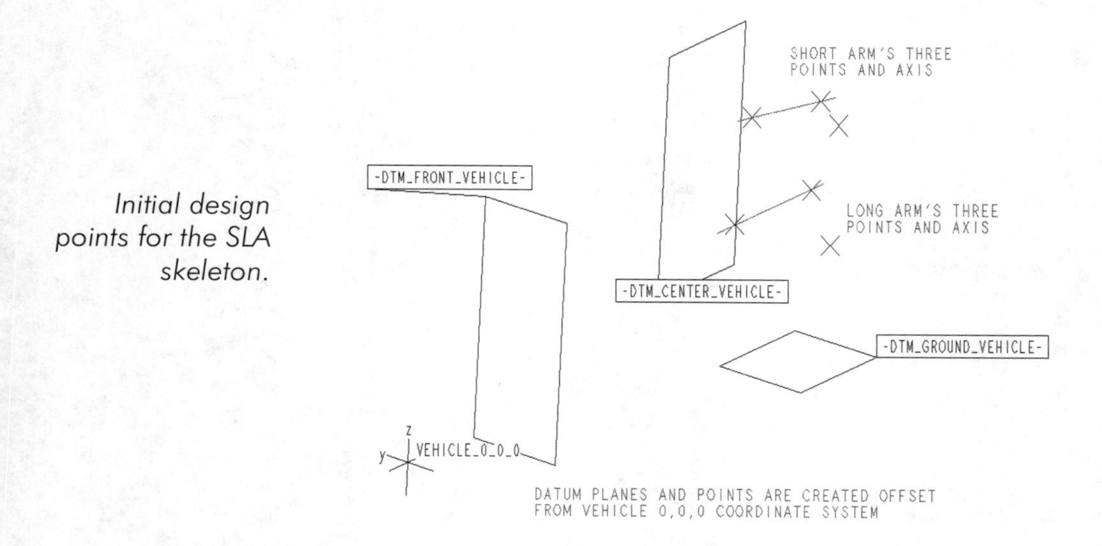

Initial design points for the SLA skeleton.

Long Arm Skeleton Development

The long arm (lower control arm) position will rotate about its design axis. The rotation creates a path (curve) developed through the datum point or tip of the long arm.

Once the path has been developed, a new datum point should be developed which controls the location of the tip at positions along the path. This position describes the tip or ball joint location for the long arm when it is assembled. The point is created by intersecting a datum plane that is offset from the horizontal ground plane and the long arm path curve. The offset value becomes the driving dimension for the suspension.

Path of rotation for the long arm.

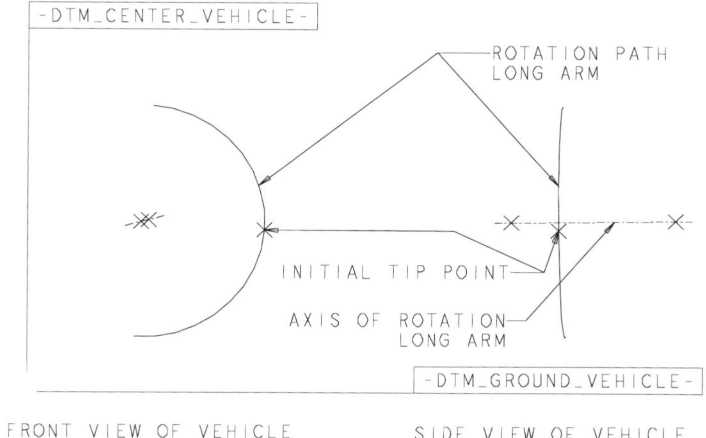

Long arm in vehicle position.

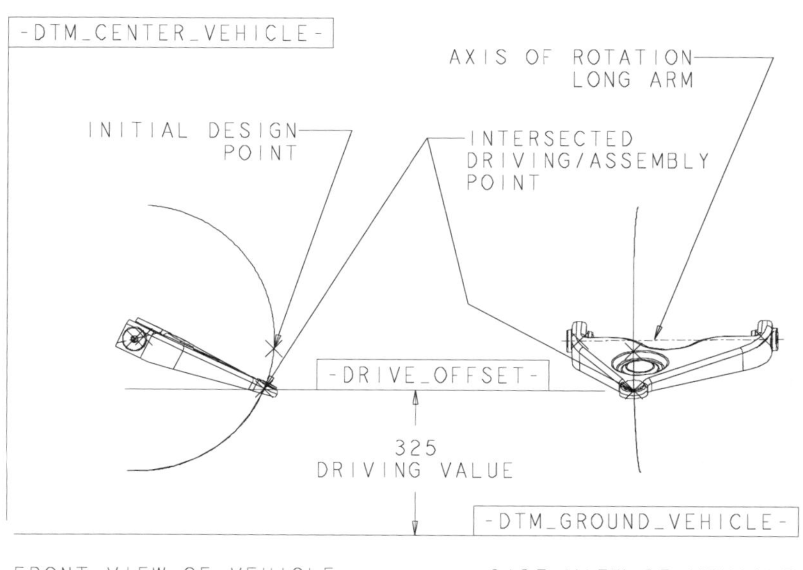

✔ **NOTE:** *The dimension offset which controls the tip or ball joint location of the long arm will be the driving dimension for the kinematics*

applied to the rest of the skeleton. Take the time to properly name important features as well as change the symbolic dimension values to meaningful names (e.g., D25 to DRIVE_OFFSET).

Short Arm Skeleton Development

The short arm (upper control arm) position will also rotate about its design axis. The rotation creates a path (curve) developed through the datum point or tip of the short arm.

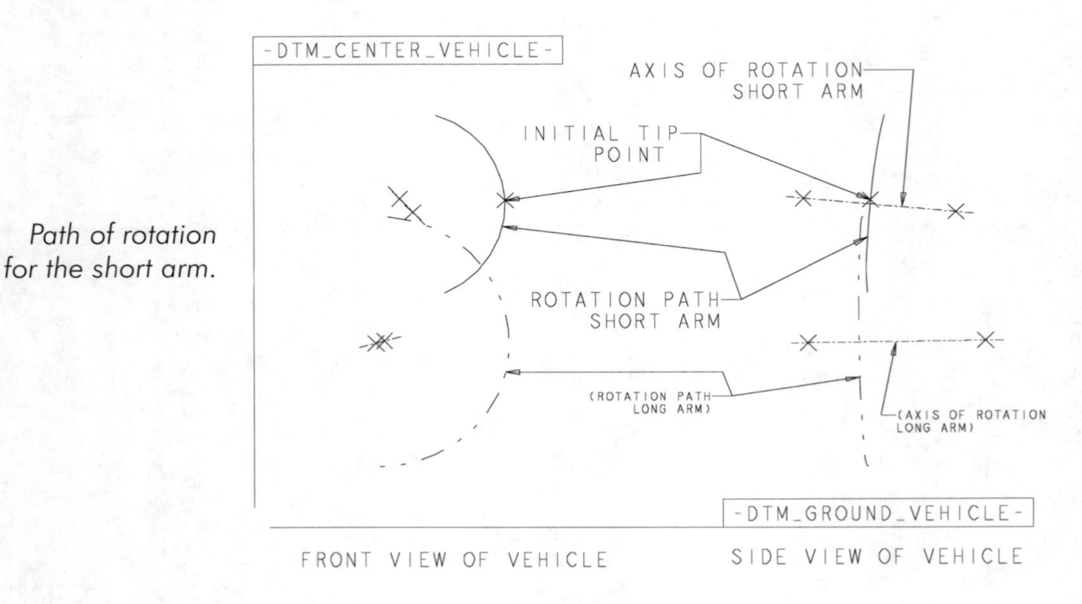

Path of rotation for the short arm.

The tip or ball joint location for the short arm will be controlled through a fixed-length dimension developed from the tip or ball joint location of the long arm. The value for the fixed length is derived through the creation of a Datum Evaluate feature. The Evaluate feature is a measurement from the original design points describing tip or ball joint locations.

The fixed-length Evaluate feature.

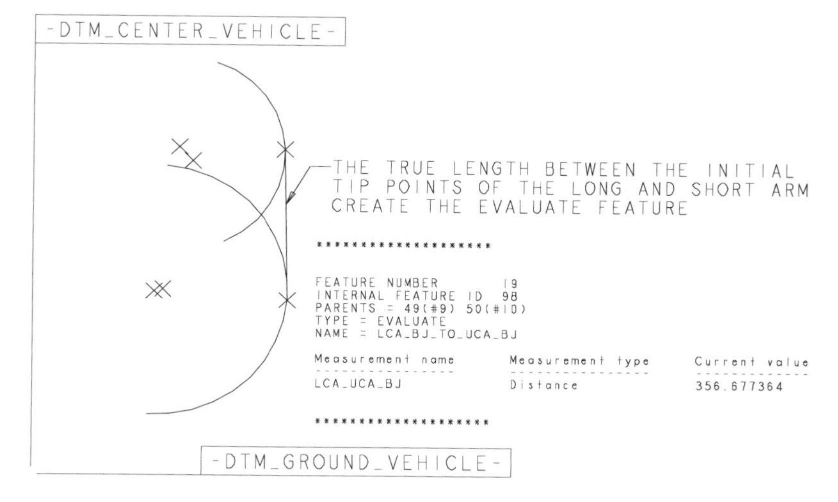

Once the fixed-length Evaluate feature is created, its value will be used to create a revolved surface with a radius equal to that value. The revolved surface is designed by attaching it to the tip or ball joint of the long arm. The surface will move along with movement of the long arm tip or ball joint. The revolved surface is then used to create the tip or ball joint location of the short arm by intersecting the surface and the path of the short arm.

Short arm in vehicle position.

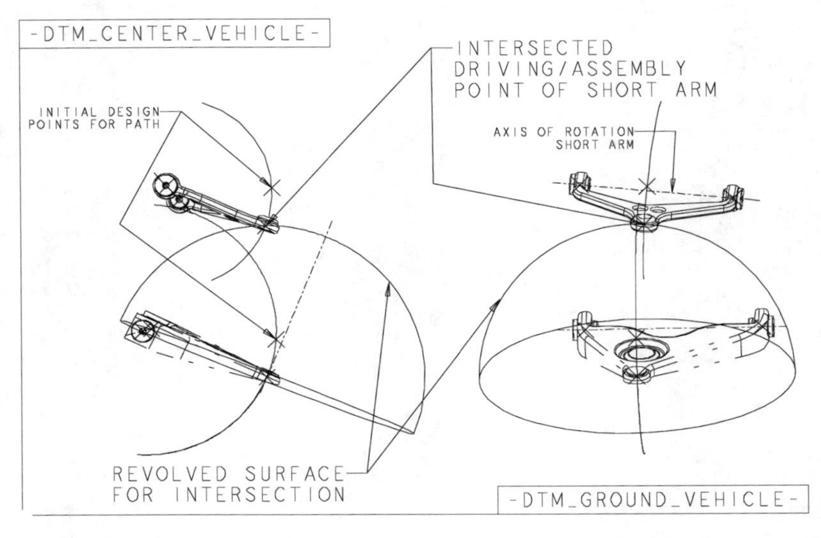

Adding the Knuckle (Fixed-length Component)

The knuckle (part between the short and long arm) is based on the design fixed length between the short and long arms. The knuckle is connected directly to the moving tips or ball joints of both the long and short arms. The third point used to assemble the knuckle is an additional point. Steering or turning the automobile is the knuckle's function. Therefore, the third point created for assembling the knuckle must also move.

Techniques for creating the moving ball joint location of the short arm are similar to those used to develop the knuckle's movement. A path is defined for the rotation of the knuckle, and then a surface intersects the path to develop the third point. The only difference is that the surface used to intersect the path of the knuckle is linked to an additional driving point named *Rack_Travel_Turn_Ctrl* rather than to the short or long arm. The separate driving point allows the SLA suspension to move left and right independent of up and down movement.

Knuckle in vehicle position.

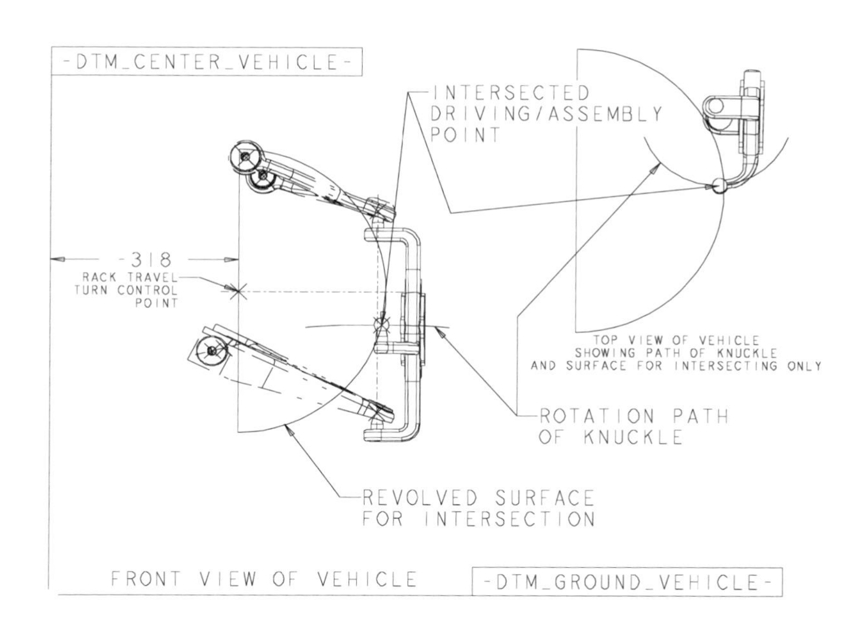

Verifying and Testing the Kinematics Assembly

Our objective thus far was to create a three-dimensional kinematics assembly. To verify that the parts move correctly, changes to the skeleton part should drive the completed assembly to any desired kinematic position.

To begin testing, modify the skeleton part in the assembly. Select the following commands:

Modify → Mod Part → Sel By Menu → Skeleton Part → Sel By Menu → Drive Offset

Enter a new test value. Regenerate only the skeleton part by selecting Regenerate while in the Mod Part menu. This regeneration is the most efficient because Pro/ENGINEER will only recalculate the skeleton part and reposition the components. Repeat this process until satisfied with the assembly function.

The completed assembly.

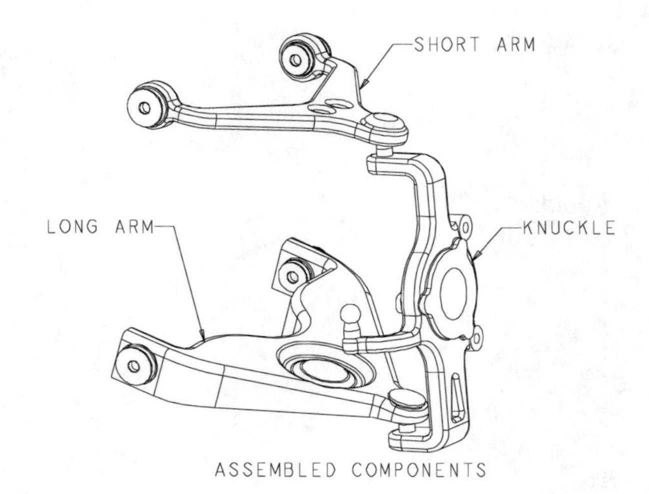

Testing positions of the skeleton part and assembly.

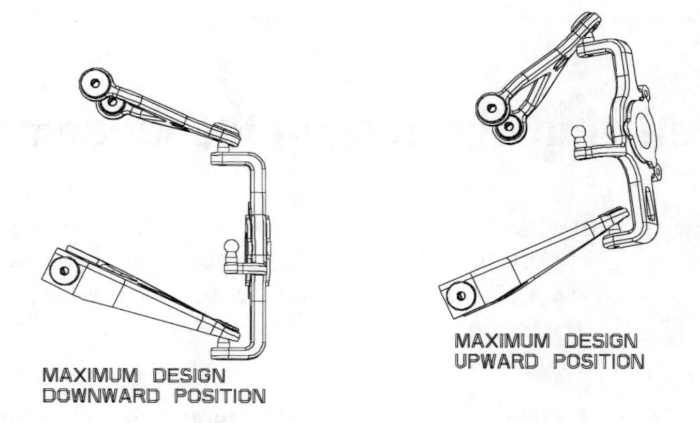

Automating the Kinematics Assembly

To take the design and function of the assembly one step further we will automate the SLA suspension by creating a set of relations and controlling them through Pro/PROGRAM. This program will increment the value of the Drive Offset parameter every time the skeleton part is regenerated.

Select Edit Design when using Pro/PROGRAM, and add the following program to the relation section of the file. This program will create the parameters for a lower and upper location and increment the Drive Offset value until a maximum or upper location value is reached. Once the upper location value is reached, the program will automatically reset itself back to the lower location value.

```
RELATIONS
/* MOTION PARAMETERS
LOWER_LOC = 325
UPPER_LOC = 575
NUM_INCREMENTS = 10
/* CALCULATION OF KINEMATICS DRIVER
INCREMENT = (UPPER_LOC - LOWER_LOC)/NUM_INCREMENTS
DRIVE_OFFSET = DRIVE_OFFSET + INCREMENT
/* RESET KINEMATICS CONDITIONS
IF DRIVE_OFFSET => UPPER_LOC
 DRIVE_OFFSET = LOWER_LOC
ENDIF
END RELATIONS
```

✔ **NOTE:** *Parameters created using Pro/PROGRAM can be easily changed by selecting* **Setup → Parameters → Modify** *in the skeleton part.*

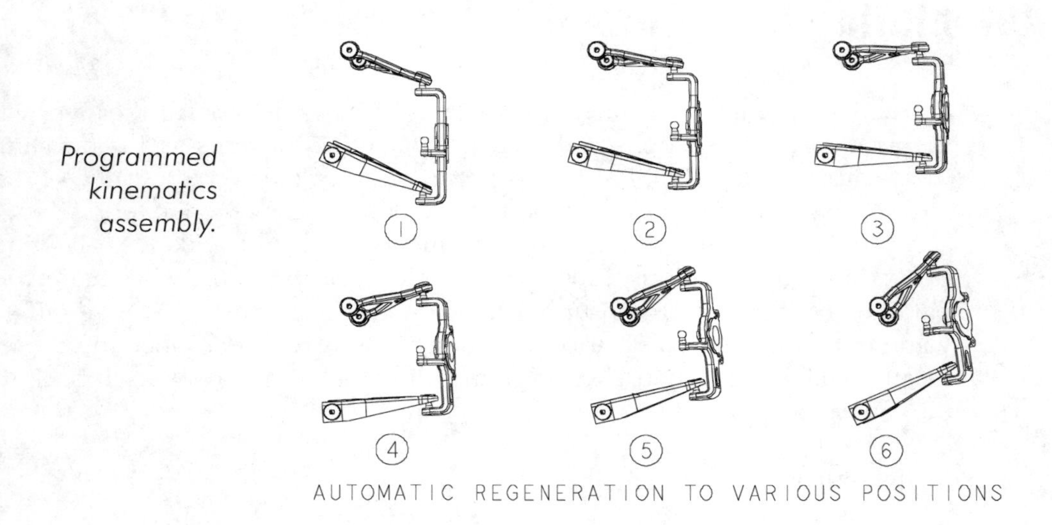

Programmed kinematics assembly.

Summary

In this chapter we illustrated that the common requirement for assemblies that move can be accomplished using the skeleton part technique. Kinematics can be accomplished in two- and three-dimensional space. The skeleton part technique provides maximum efficiency to manipulate, study, and test assemblies. The skeleton technique also provides the ability to manage large assemblies by allowing for disassembly and representations to be made. We also demonstrated how to automate the assembly, thereby providing another tool for going beyond the basic design.

Working with Large Assemblies

Introduction

Managing large assemblies in a workable fashion has been a concern of many Pro/ENGINEER users. Performance and productivity are of the utmost importance. However, advanced users often find that their assemblies contain so much information that current hardware or traditional software tools inhibit performance. Pro/ENGINEER has addressed these issues with multiple tools for the engineer and designer to enhance the product development cycle. In this chapter we discuss the use of various tools to improve performance and productivity in large assemblies. We briefly mention some important traditional tools that still help performance. In addition, more advanced tools such as Simplified Representations using Included and Exclude, By Rule, By Envelope, and By Representation are examined in depth.

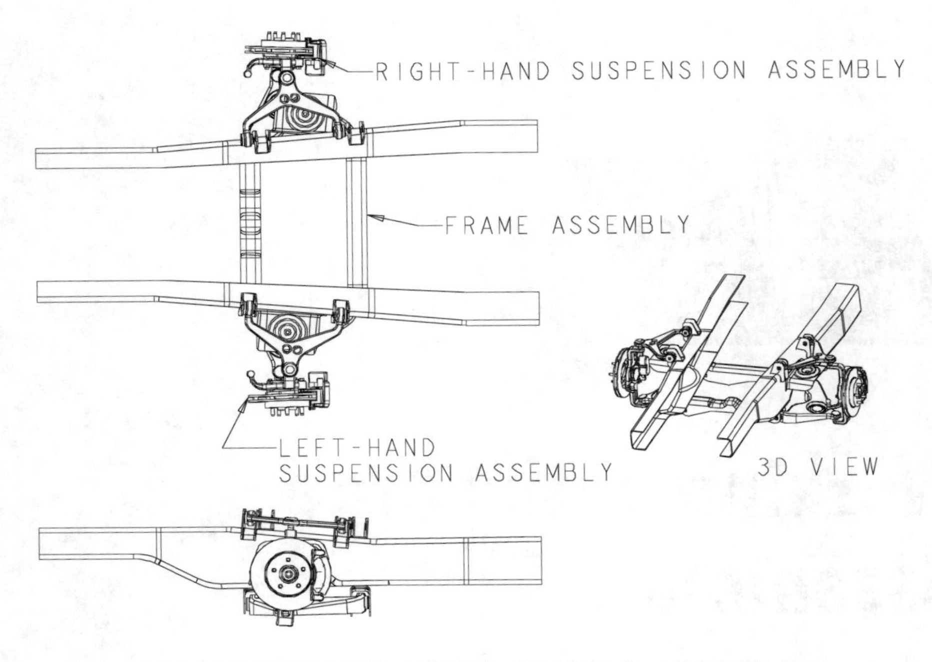

Example of a large assembly: automotive front chassis assembly (99 parts).

The Advanced User's Basic Guidelines

As you develop large assemblies containing other assemblies with complex parts, performance will diminish. The reduction in performance can be improved by using the following basic techniques:

- Take advantage of layering at the part level. Layering can greatly improve graphical performance in a number of ways. In large assemblies, you can create layers at each part level so that items you do not need—such as construction curves, points, and datum planes—can be easily Blanked.

- Take advantage of Layering at the assembly level. Layering components of an assembly can also greatly improve graphical performance in a large assembly. Layering each component allows the user to easily Blank or Display components which are necessary only for the current task.
- Work with geometry displayed as wireframe. This simple tip can save a tremendous amount of time because it removes the hidden line calculations.
- Suppress components that you do not need. If the assembly has logically developed parent/child relationships, some components can be suppressed temporarily without compromising design goals.
- Whenever possible, work with skeleton parts as described in Chapter 14. Skeleton parts allow for many design iterations when studying movement. They also allow for flexibility to be tested before entering the assembly.

The above techniques establish some of the basic steps that an advanced user applies when beginning to work with a large assembly.

Advanced Large Assembly Management Tools

Using Simplified Representation (Include | Exclude)

One of the most significant improvements for working with large assemblies was the introduction of Simplified Representation (originally named Configuration States). This enhancement provides instant access to any level of the assembly hierarchy structure.

✘ **TIP:** *To view the hierarchy structure of an assembly at any time prior to entering Simplified Representation, select* **Info → Assy Tree**.

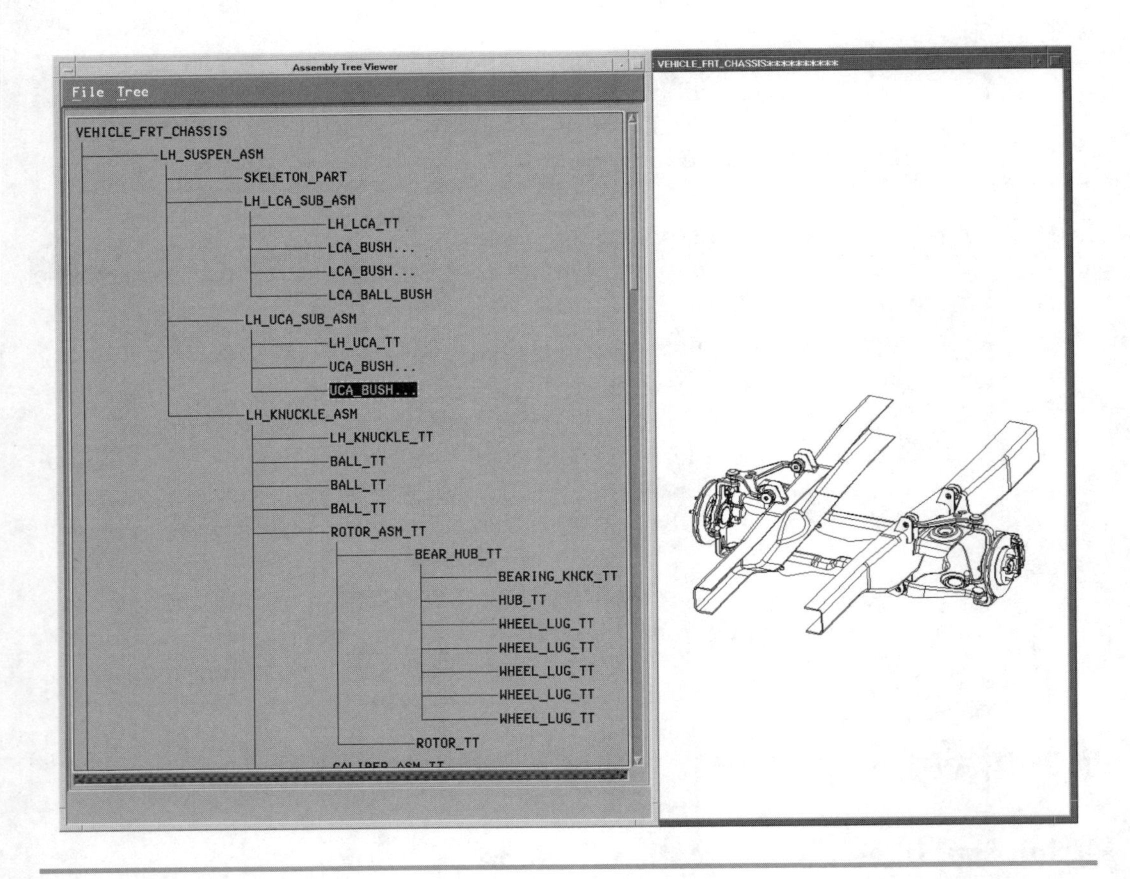

Assembly Tree of automotive front chassis.

Directly entering Simplified Representation enables you to easily include or exclude components immediately affecting the performance of a large assembly. All simplified representations are based on a "master representation." The master representation contains all components of the assembly. Select the following:

Simplified Rep → Create → (enter name of representation, i.e., FRAME) → Include (or Exclude)

Next, select the assemblies or parts using Editor Pick (the default, selecting from the Component Editor window), Screen Pick (selecting the assembly or part on the screen), Range, or All, or by using Rule, Envelope, or Representation. The latter three methods will be examined later.

At this point, you would select **Next Update Rep → Done**. This action sets the assembly to the most beneficial viewing configuration to continue the design. For example, upon selecting the frame assembly to include from the Component Editor window, the large chassis assembly is quickly reduced in complexity to a few parts.

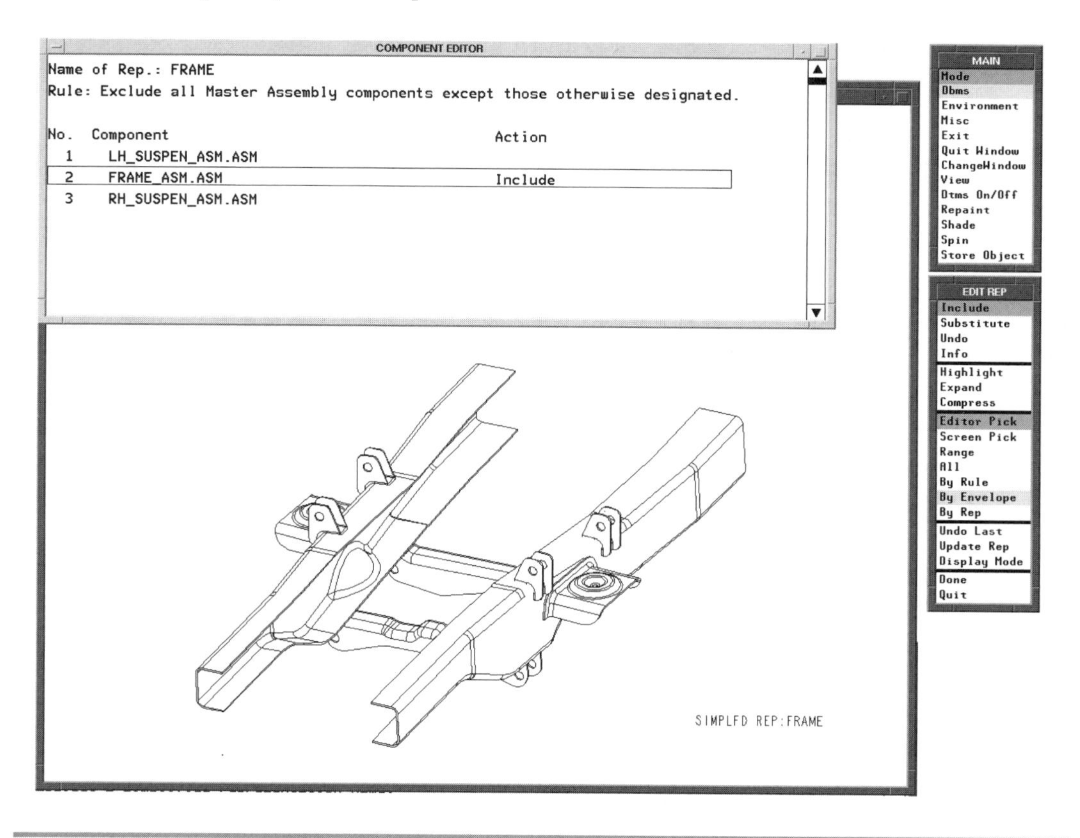

Simplified Representation showing the frame assembly.

✗ **NOTE:** *When you retrieve an assembly, Pro/ENGINEER automatically brings all parts and respective assemblies into the current working session (i.e., they are loaded into RAM). When you use Retrieve Representation, you have the ability to bring only the representation's* *components into the current working session. This technique will also greatly improve performance because of the reduction in RAM required. However, this reduction in RAM only occurs during the initial simplified retrieval. Parts added to the representation later will be incorporated in the current working session (loaded into RAM).*

While in the Simplified Representation, you can access all Pro/ENGINEER functions with the following exceptions:

- Restructure, Family Tab, and Integrate options.
- Delete or Suppress Substituted components.
- Redefine components that have been Excluded or Substituted.
- When creating a drawing, you cannot make an Exploded view of a Simplified Representation.

Expanding Simplified Representations Using Rules

As mentioned earlier there are a number of ways to select the components to Include or Exclude in a simplified representation. Selecting components using the By Rule command allows you additional powerful options to develop simplified representations. The By Rule option offers component selection through the use of Model Name, Size, Distance, Expression, and Zone.

Using By Rule - Model Name

Selecting components by using the **By Rule → Model Name** option is an extremely fast method to establish a representation of similarly named parts. In the large chassis assembly there are a number of components (bushings) that are used to help anchor the suspension components to the frame. To create a Simplified Representation showing only the bushing components, Model Name is used. During product development this tool provides easy access to similarly named parts for design verification and changes. This technique for selecting parts first reviews the assemblies. If a match does not occur, each component is automatically reviewed.

TIP: *When using the Model Name option, you are prompted to enter the model name to search for. This option allows for the use of wildcards (e.g., *bush* to select all of the bushings).*

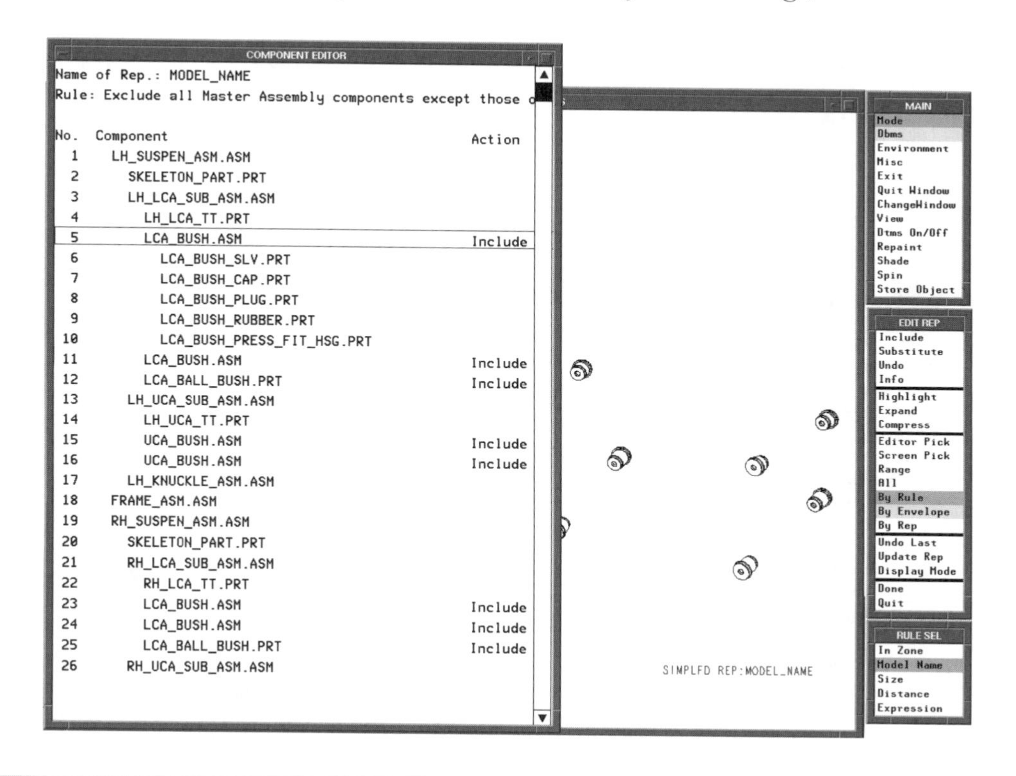

*By Rule - Model Name *bush*.*

Using Rule - Size

When using **By Rule → Size**, you must understand how Pro/ENGINEER determines size. Size is determined for every part and assembly by the parameters of the smallest box that fully encloses the part or assembly in three-dimensional space. This "bounding box" includes all datum planes and surfaces. If you are working in a Simplified Representation, the bounding box reflects the current state of the parts and assemblies when determining size. Therefore, to accurately calculate the size, the master representation should be used when creating a representation using Size.

After selecting Size, you must then select from two other option sets: Relative or Absolute, and Greater Than or Less Than. Relative size prompts you to enter a size limit relative to the size of the top-level assembly. This size is a value between zero (0) and one (1), where 1 is a bounding box containing the complete top-level assembly. Absolute size prompts you to enter a bounding box size comparable to the units of the top-level assembly. Greater Than is used to select models greater than the size limit entered. Less Than is used to select models less than the size limit entered.

The example illustrated below shows **By Rule → Size → Absolute | Less Than** (size entered is 500). Each component shown represents a bounding box with a size less than 500 units.

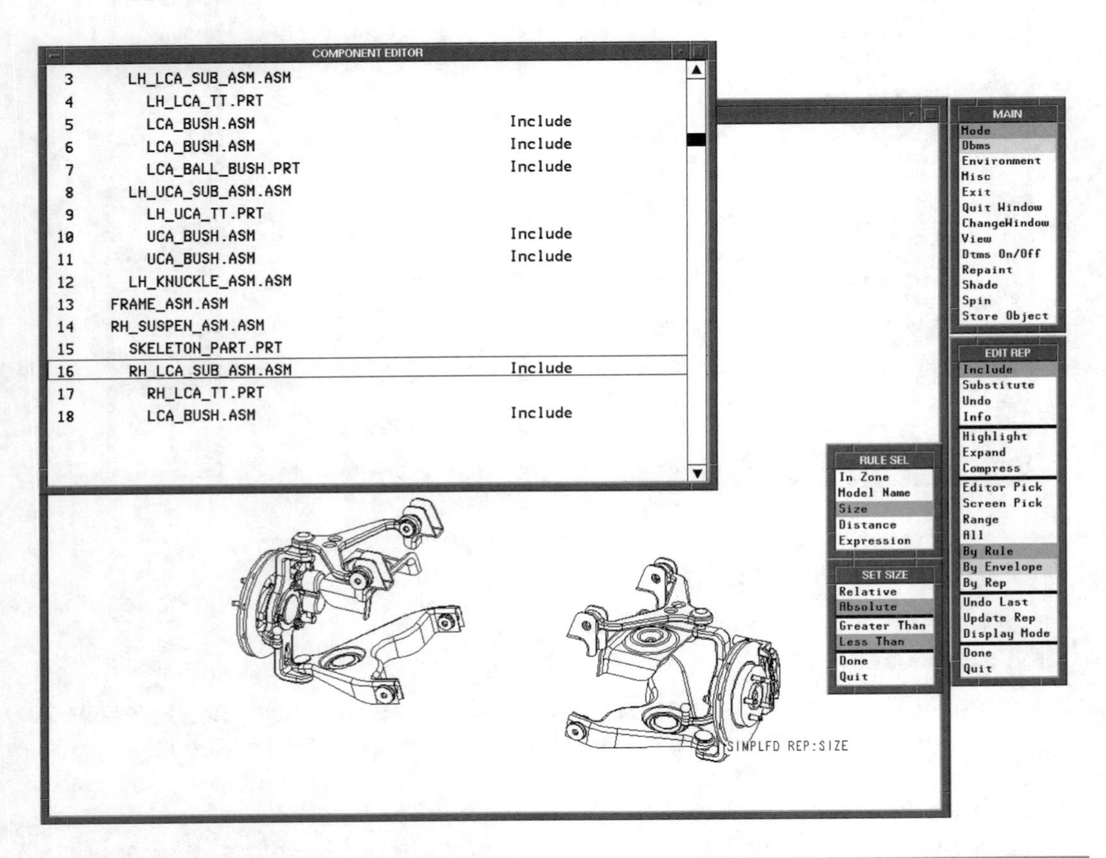

By Rule - Size.

✘ **TIP:** *The bounding box for a component's size is stored with the meta data. This means that Pro/ENGINEER has fast access to the size information. The information can then be used during retrieval (Retrieve Rep) of the assembly in a simplified representation, thereby saving RAM and improving performance.*

Using By Rule - Distance

Selecting components using **By Rule → Distance** offers yet another tool to create a Simplified Representation. The function of the Distance option is quite simple. Distance allows you to select a Point/Vertex, Component Center, On a Surface, or Offset from a coordinate system to develop a sphere. The sphere created by the option has a radius of the distance value that you enter. Any component inside or intersecting the sphere can then be Included or Excluded. This rule is most often used to quickly develop a design area inside the sphere. For example, in the large chassis assembly the designer may wish to focus on the area near the brakes. Selecting **By Rule → Distance → Point/Vertex** and picking a vertex near the brake with a distance of 25 creates that design representation.

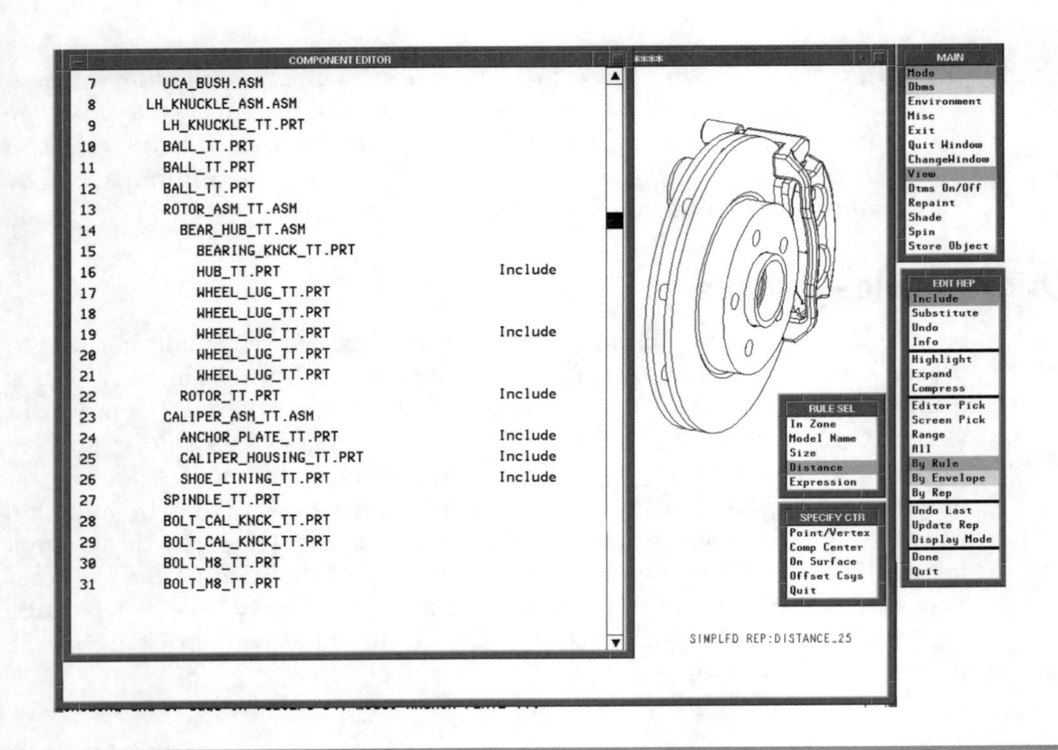

By Rule - Distance (near left-hand brakes).

Using By Rule - Expression

Using **By Rule → Expression** is another effective way to select components in Simplified Representation. Upon using Expression, you are prompted to enter a logical expression containing parameters from assembly components. A good example is to assign a parameter to components that equates to "cost." The Expression could then be entered as *COST >= 25*. By Rule would then select the parts that have been assigned a cost parameter of 25 or higher.

✘ **TIP:** *An Expression can contain multiple operators. For instance, using the above example, the expression could be extended as follows: COST >= 25 & TYPE = FRAME.*

Using the By Rule Zone Option

Using the **By Rule → Zone** option inolves additional setup. A Zone can be best described as an area developed on one side of a datum plane. This Zone (area) can be used in a Simplified Representation to select all parts Included or Excluded in the Zone.

Before a Zone can be used in Simplified Representation, it must first be created. The creation of a Zone is accomplished by selecting **Assem Setup → Zone → Create**. Next, enter the name of the Zone (e.g., *Right_Side*). Pro/ENGINEER will then prompt you to Select or Create a Datum Plane to relate the Zone to. You will have the option to choose which side of the Datum Plane the Zone will be created on. Using the Flip command allows you to change sides for the Zone. Once the Zone is created, select **Okay → Done**.

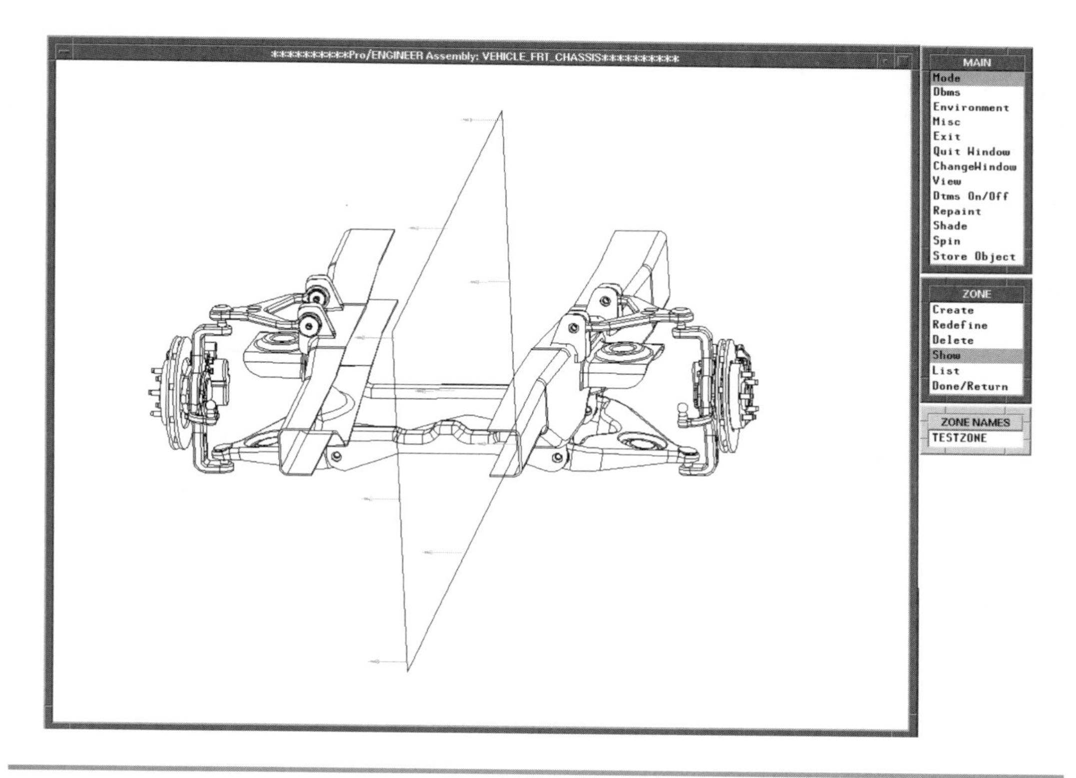

The Right_Side Zone.

Multiple Zones can be created. Using the Show command while in the Zone menu will graphically display each of the zones you have created.

Using a Zone in a Simplified Representation is very effective for quickly selecting areas of components. While in Simplified Representation, select **By Rule → Zone**. A list of all created Zones will appear. Select the Zone to Include or Exclude components (e.g., *Right_Side*).

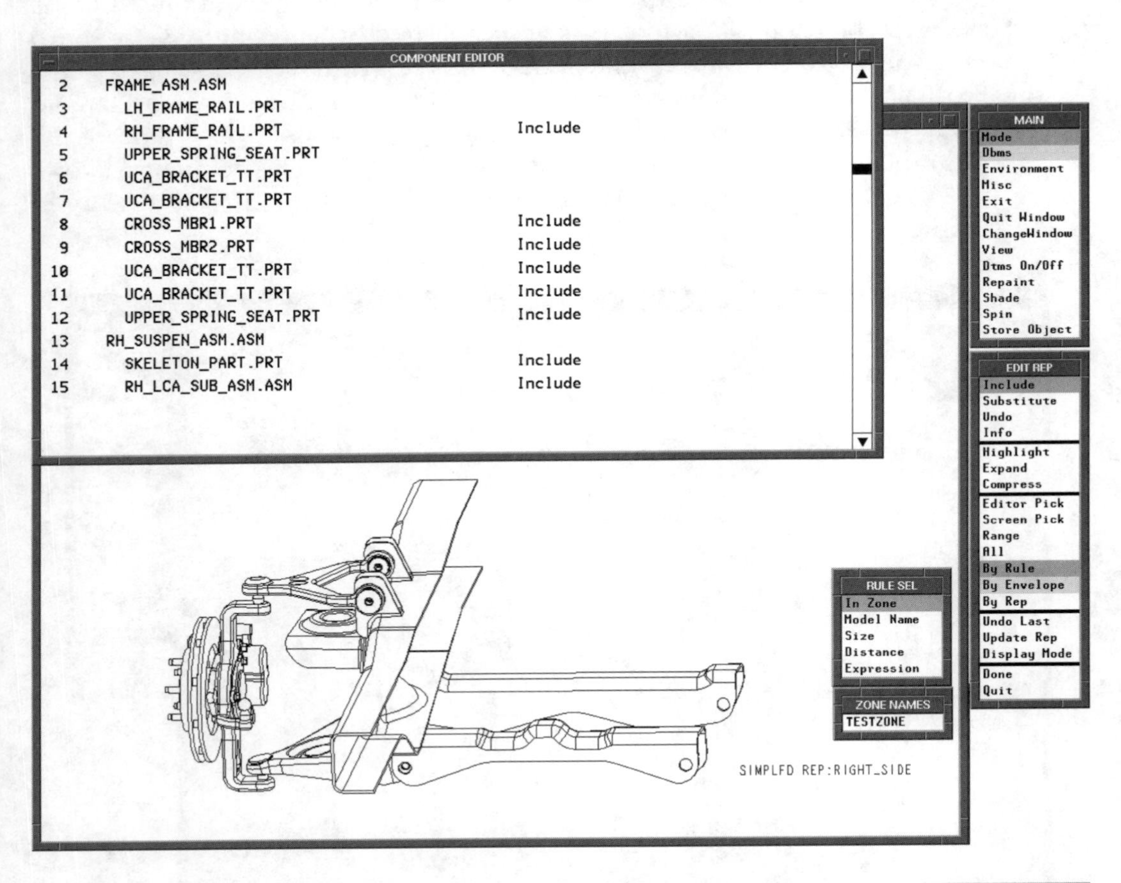

By Rule - Zone. The right side of the large chassis assembly.

Additional Options in Simplified Representation

For the advanced user to take the above tools one step further, three other options must be examined. These options include the creation of Envelopes as well as using Substitute and By Rep in Simplified Representation.

Creating an Envelope

While working in a large assembly, you may often need to view portions of other individual parts or assemblies without viewing their complete complex geometry. The Envelope tool allows the designer to capture any portion of existing components by creating a single part representing the desired portions. The Envelope can then be used in a Simplified Representation as a Substitute for the components it represents. To illustrate how this tool works, we will use an example.

Assume that a design change requires changes on a left-hand suspension assembly, which in turn must be implemented in the large chassis assembly. The complete frame assembly is not necessary to make the changes, although portions of the frame assembly are still important. The surfaces around the mounting locations for the suspension should be visible while making the changes. To establish a simple version of the complete frame assembly, an Envelope part is created. Select **Assmy Setup → Envelope → Create** and enter the name of the Envelope (e.g., *ENV_1_FRAME*). Next, the Sel Membrs menu and the Component Editor window appear. Select the complete frame assembly for the Envelope. Select Create and enter the name for the Envelope part (*FRAME_ENV*). Next, create the new part which will represent the desired components.

Many advanced users will create the Envelope part using Pro/ENGINEER's Surface Copy function. Surface Copy allows you to capture representative geometry in a simple form. For the frame assembly, the surfaces around the mounts are copied.

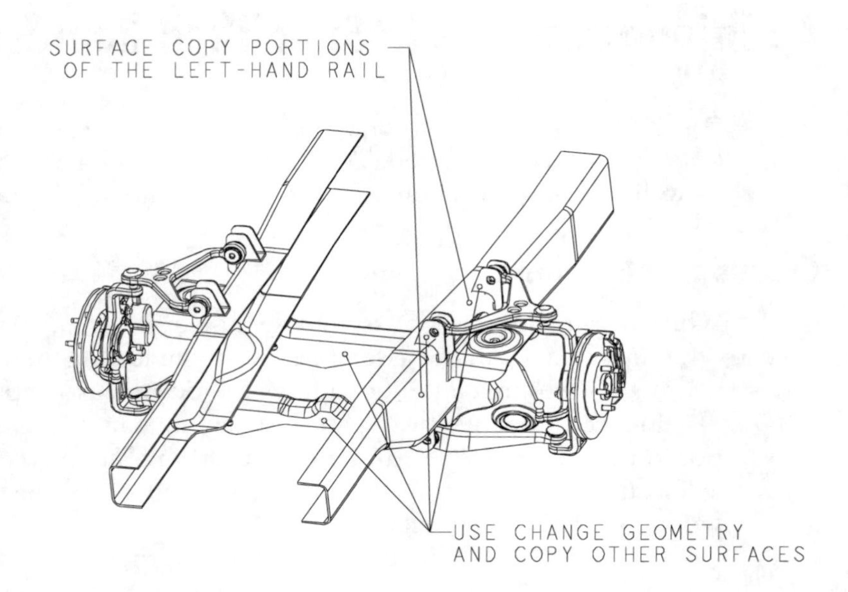

Using Surface Copy to create an Envelope part.

Once the part is complete, changes can be easily made to the Envelope part using the **Redefine → Change** option in the Envelope menu. The Change menu includes the following options:

- Names—Rename the envelope independent of the part name.
- Members—Add or remove the list of components in the Envelope.
- Geometry—Modify the geometry of the Envelope part (e.g., adding Surface Copies).
- Part—Select parts to be used as the Envelope part. Displays the Env Comp menu.

In addition to Redefine in the Envelope menu other options allow for Deletion, Visibility (turns on display of the Envelope part), Show (temporary display of the Envelope part), List (lists representations the Envelope part is used in), and Info (complete information on the Envelope part).

✔ **NOTE:** *Envelope parts are stored as part files in .prt. However, they are only accessible from the assembly in which they were created.*

Using the Envelope Part as a Substitute

After an Envelope part or parts has been created, you can directly use the same in a Simplified Representation. Selecting Substitute "un-grays" the By Envelope option. Choosing By Envelope will give you a list of available Envelopes. Selecting the appropriate Envelope automatically replaces the Envelope members (e.g., the frame assembly) with the "single" Envelope part. This simple substitution reduces the number of components in the assembly as well as the number of features in the parts substituted. In our simplified representation, we will also choose to include the left-hand suspension assembly.

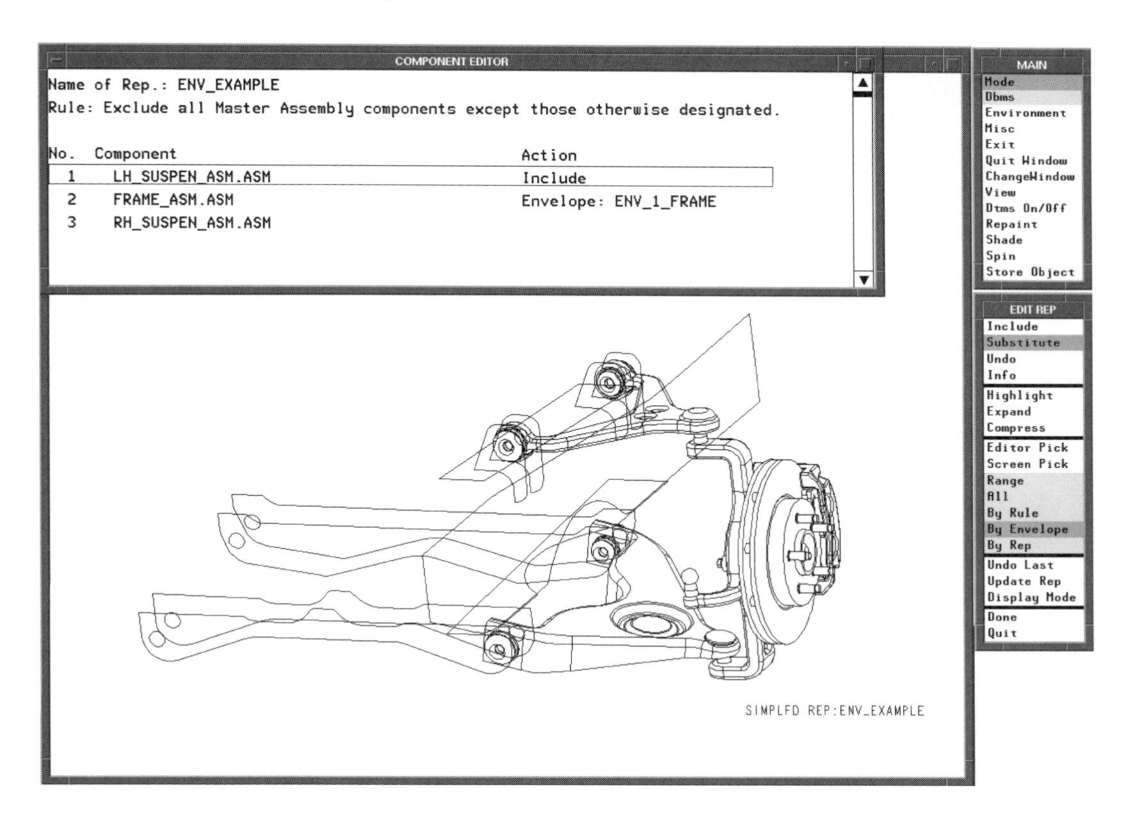

Envelope part and left-hand suspension for design changes.

✗ **TIP:** *The Substitute option is not limited to Envelope parts. Substitute can also use Family Table members, Interchange Groups, and Simplified Representations of objects.*

Using Other Representations in Simplified Representations

This function is often overlooked by advanced users. Once various representations have been created, you can access these representations while creating or redefining a new simplified representation. Selecting **Include** or **Exclude → By Rep** and choosing from the existing representation automatically develops new combined representations.

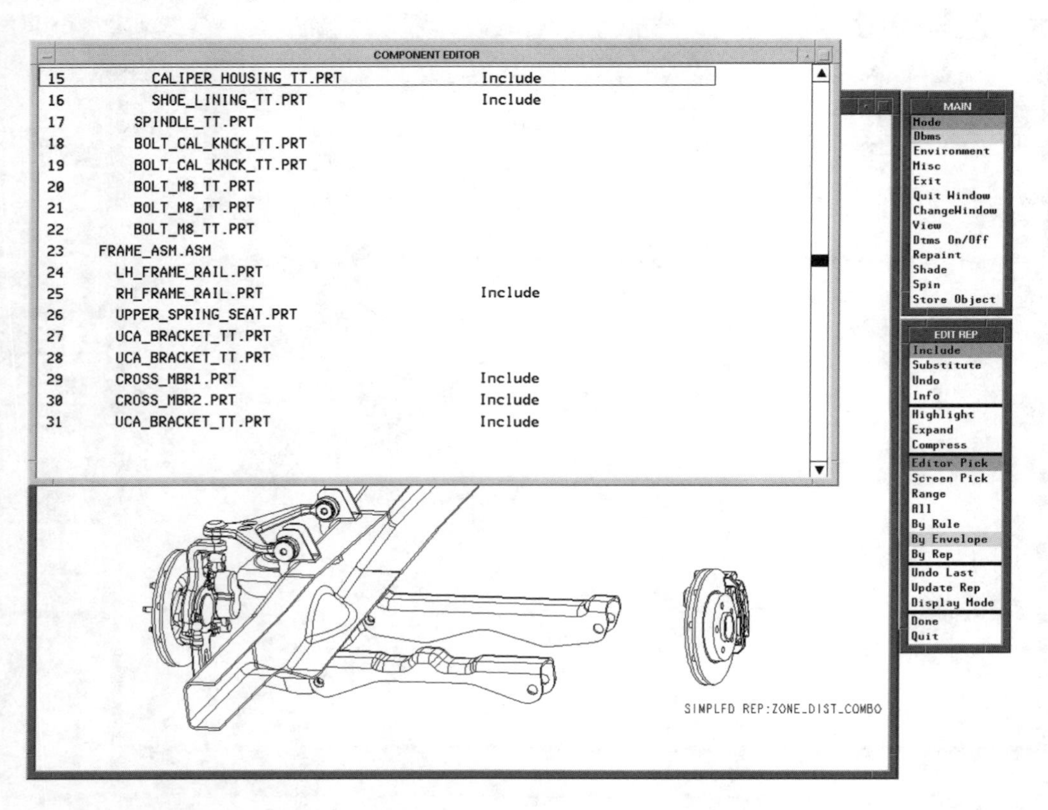

Combining representations, zone, and distance in a new simplified representation.

This function allows for quick improvements or development of a simplified representation. For example, a new simplified representation could be made to include both representations from the zone and distance examples.

Summary

The management of large assemblies in a workable, efficient fashion is greatly improved through the use of multiple tools within Pro/ENGINEER. The tools examined in this chapter illustrate how you can directly enhance performance in large assemblies. Using a basic technique such as working in wireframe display or developing Simplified Representations by Rules or with Envelope parts supplies you with the ability to improve performance and productivity.

Master Modeling

Introduction

Master modeling is a technique used to spawn multiple associative derivative models. This method involves incorporating a merging process of the master model into individual component models. The unused area of the component model is removed, leaving the desired portion of the component model. The major advantage in using this process is that changes made to the master model are automatically reflected in the derivative models.

Master modeling is particularly useful when trying to match key features and complex surface continuity between individual assembled components. Candidates for this method are ergonomically designed plastic components such as computers, telephones, CD players, telephones, copiers, and so forth. The most important aspect of this process is planning the features to be contained in the master model that relate to the individual component models.

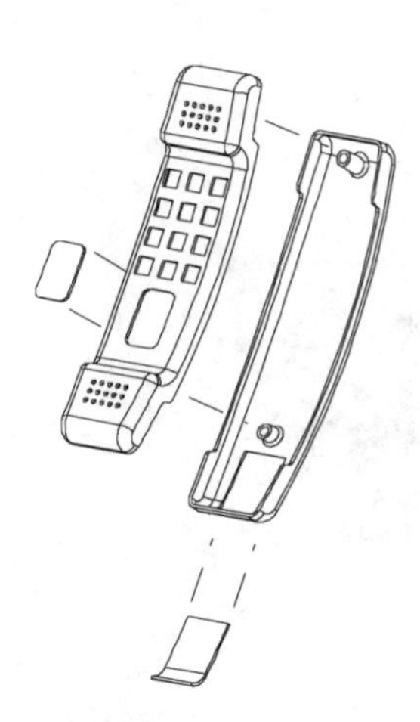

Telephone assembly.

Process Overview

When designing for assembly, many factors must be considered in determining how well an assembly fits together. Up-front planning is required in order to focus on key issues such as parting lines, draft angles, tongue and groove joints, mounting bosses, design tolerances, and overflush or underflush conditions. These issues must be considered in the planning of the feature content of the master and derivative models. The master model must contain not only the global definition of the assembly, but key reference features such as points, planes, curves, and surfaces. These key datum features are referenced in the derivative models for sectioning the models into individual component models. This is accomplished by creating a temporary assembly model and merging the master model into each subcomponent model. Unwanted geometry is removed by referencing the globally defined parting geometry in the master model.

Description of Master Modeling

Appearing below is a brief step-by-step description of the master modeling process.

1. Create a master part model that globally defines the assembly envelope of space desired.
2. Define key points, planes, curves, and surfaces used for sectioning the model and the subcomponent level.
3. Create the potential derivative part model consisting of default datum planes only.
4. Assemble the derivative model into a temporary assembly without creating any assembly features.
5. Assemble the master model directly into the derivative part model, and reference the default datum planes of both models.
6. Merge, by reference, the master model into the derivative model using the **Component/Adv Util → Merge** command. Detach the master model from the assembly when prompted.
7. Repeat the steps as desired for as many derivative models as required by the design outline.

✘ **TIP:** *Once the master model has been merged into the derivative model, use a copy save to quickly create other derivative models.*

8. Delete the temporary assembly because it is no longer needed.
9. In each derivative model, create a layer that contains the external reference geometry so as to display the parting geometry when required.
10. Complete the model by creating additional finishing features such as shells, flanges, ribs, and bosses. Make sure to reference the master model reference features.
11. Create a finished assembly and check for form, fit, and function across mating surfaces.
12. Flex the master model at key critical areas to check for derivative model associativity.

Tips and Techniques

- When creating mating features, such as common lips, edges, or flanges, always reference features from the master model.
- Next, define sweeping curves in the master model to ensure commonality between individual component models.
- When merging the master model into the derivative models, do not copy datum planes because they should already have been defined.
- It is critical that the temporary assembly model *not* contain any assembly features. The temporary model creates external references used for managing the associativity between each component. The goal is to create the direct link between the master model and its derivatives without carrying the extra baggage of an intermediate assembly.
- Include the Shell feature in the master model when common wall thickness and draft features are desired between the mating components.

Telephone Assembly Example

In this section, we will develop the telephone assembly example.

Exploded view of telephone assembly.

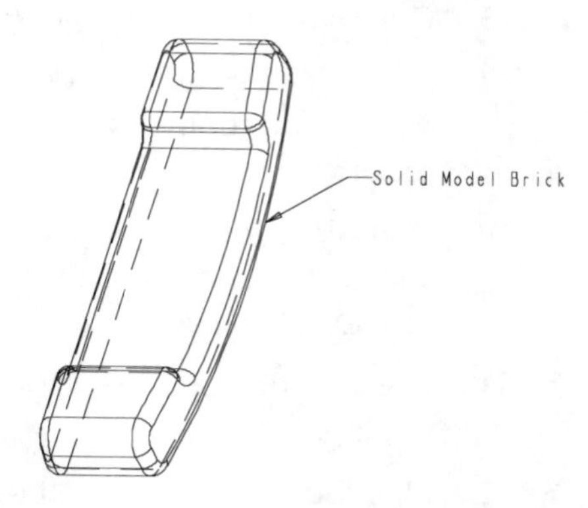

First, create a solid model representation of a telephone which defines the outer surfaces of a finished assembly.

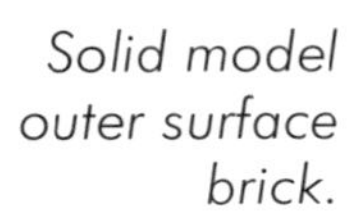
Solid model outer surface brick.

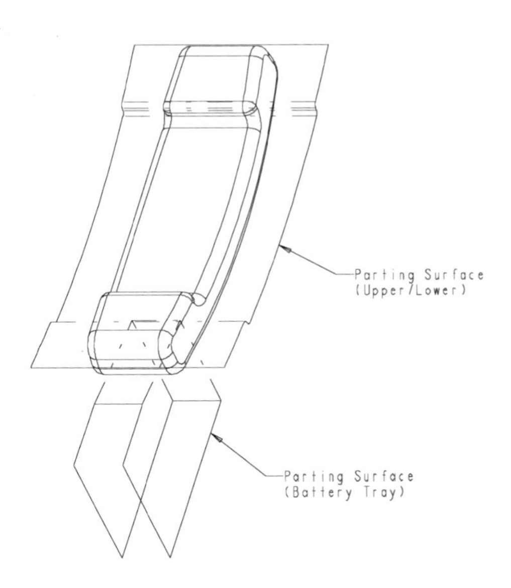

Develop the parting surfaces that represent the mating interface between the upper and lower halves. Define the parting line for the battery tray and pushbutton by sketching the shapes on a plane and projecting on the appropriate surfaces. Add additional features such as the location for mounting bosses.

Parting geometry.

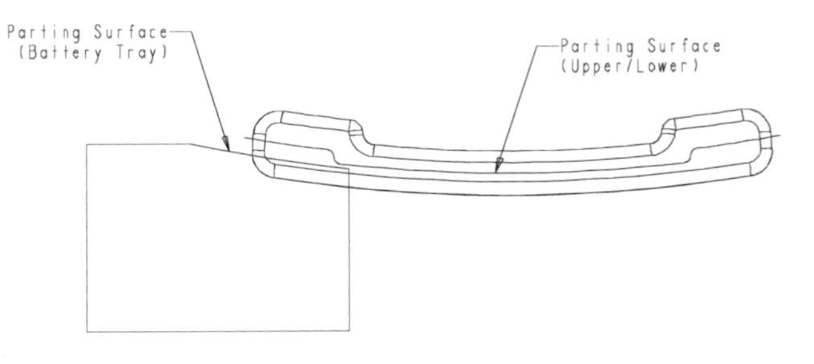

Subsequent steps in the telephone assembly follow:

1. Create a new part called *upper* which contains the default datum planes.
2. Assemble the *upper* part to a temporary assembly which contains no assembly features.
3. Assemble the master model while referencing the default datum planes or coordinate system from the *upper* part.
4. Merge the master model part into the upper part using the following command sequence: **Component → Adv → Util → Merge → Refernec**.
5. In the *upper* part, review the copied master model geometry. Create a layer called *master* and layer away reference geometry.

Master model reference geometry.

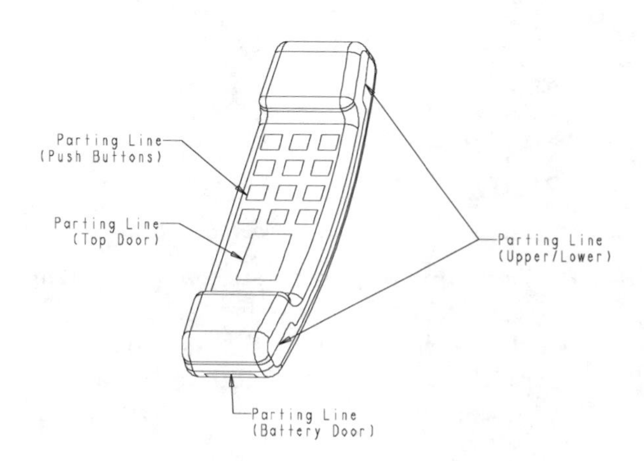

6. Copy the *upper* model into the appropriate other models (lower, battery tray, and top door).
7. In each subcomponent model, create finishing features such as shells, cuts, and bosses that reference the merged master model geometry.

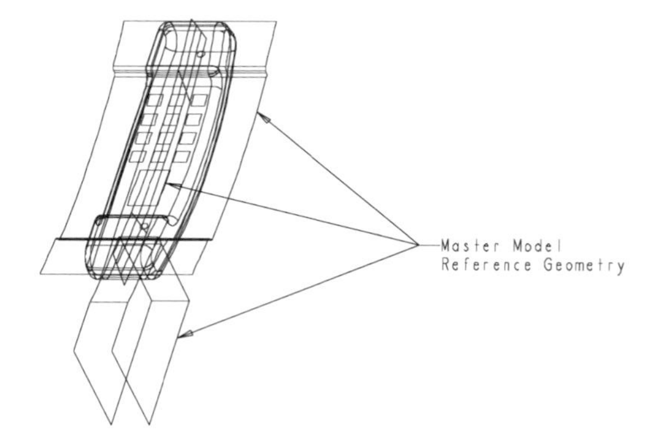

Finishing features.

8. Create a final assembly that contains all the derivative subcomponents. Check for form, fit, and function requirements. Delete the assembly used for merging the master model into the subcomponent models.
9. Flex the master model and check to make sure that the subcomponent models update accordingly. Add additional geometry to the master model if required.

Summary

The master model method is extremely effective in cases where you want to control the fit and finish characteristic of an assembly. Defining the controlling features up front in a master model guarantees the governing associative features for the derivative subcomponent parts. Additional features added to the master model always show up in the first feature of the derivative models. This ensures that the appropriate geometry is always available when adding features or refining existing features.

PART Five

Pro/NOTEBOOK

Using Layout Mode to Evaluate Designs

Introduction

The Layout mode used in conjunction with the Pro/NOTEBOOK module is often overlooked or ignored because of lack of time. In this chapter we will examine the use of Layout mode and its benefits to the engineering process and related issues. We will illustrate how advanced users can easily dive into Layout and emerge with an additional tool to improve the engineering process. The laundry detergent bottle and various caps will be used to illustrate how to develop and benefit from Layout. In addition, we will examine how to define global parameters to drive model parameters and assembly procedures (automatic assembly), and how to develop spreadsheets containing other important information.

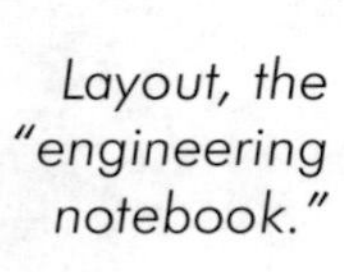

Layout, the "engineering notebook."

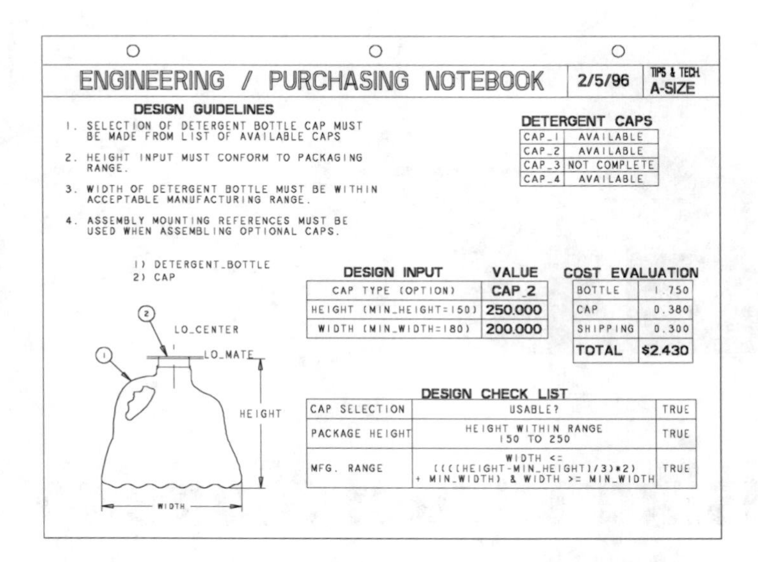

Why Layout?

The Layout functionality is often referred to as an "engineering notebook." Layout enables you to create two-dimensional sketches which can describe complex parts and their assembly requirements in a conceptual form. Although the two-dimensional sketches are not detailed they can represent your design intent, thus establishing a blueprint for the engineering process. The design intent can be shared through global parameters to components within Pro/ENGINEER. This procedure will ensure that early assembly requirements and parameters will be maintained throughout development.

Assembly requirements established in a layout can also allow components which reference (are Declared to) a layout to be automatically assembled by deriving the assembly requirements directly from the layout. Additional information can be easily added to a layout to describe critical guidelines for product development. The guidelines can be used to determine fits, sizes, and other relationships between critical design parameters. Using notes, balloons, and spreadsheets you can establish a powerful tool for development from concept to completion.

Creating a Layout

Creating a layout is similar to creating a drawing. First, select Layout from the Mode menu and proceed by entering the name for the layout. Next, the Get Format, Dwg Size Type, and Dwg Size menus are displayed. A format developed for layouts is then selected. You can can add a page by selecting **Sheets → Add** to automatically add another page with the same format.

✔ **NOTE:** *You can create and store a template format designed for a layout. The template, retrieved during the creation of the layout, should be part of your company's standards to give each layout sheet a unique, identifiable look. Typically, A size (8-1/2" by 11") is used so that the layout sheet can easily be added to a notebook.*

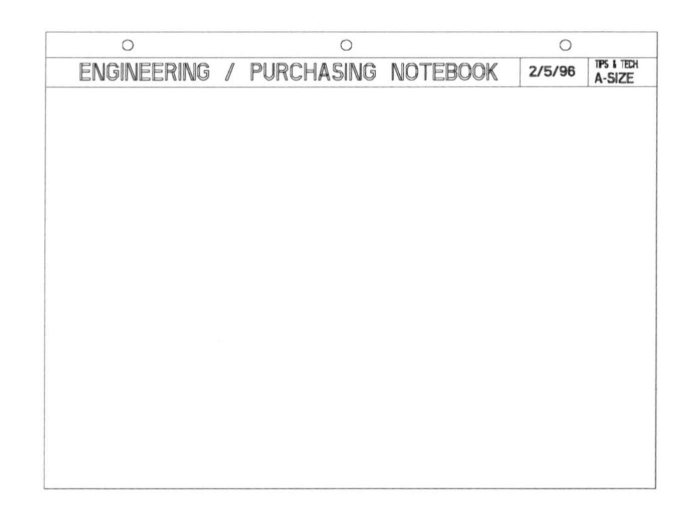

Unique format layout.

Layout Goals

The laundry detergent bottle and optional caps will serve as our layout example. Goals for the layout are summarized below:

- ❑ Two-dimensionally illustrate the components (bottle and cap) and establish assembly mounting references.
- ❑ Establish the design guidelines.
- ❑ Chart available detergent bottle caps.
- ❑ Provide an input section for critical parameters (cap type, height, width).
- ❑ Provide a design checklist to illustrate whether critical parameters meet engineering needs.
- ❑ Provide a cost evaluation chart.

In addition to the layout, additional goals will be achieved. These goals are based on the functionality of layouts to develop global parameters that can be used at the component level. The goals include the following:

- ❑ Reference the component to the layout (Declaring).
- ❑ Obtain assembly mounting information for automatic assembly of the detergent bottle and cap from the layout.
- ❑ Obtain "cap type" to determine which cap is to be assembled (see illustration) in the final assembly from the layout.
- ❑ Obtain critical parameters for "height" and "width" for the detergent bottle design from the layout.

Achieving Layout Goals

2D Illustration of Components and Assembly Mounting References

The first objective is to establish a two-dimensional (2D) representation of the components (bottle and cap). This can be accomplished by selecting **Detail → Sketch**. The Draft Geom menu will then provide a complete set of tools to sketch (2D) draft entities that represent the components.

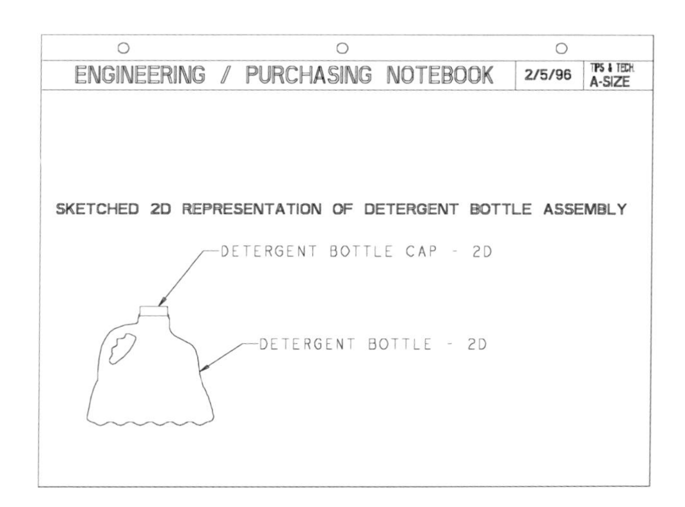

Two-dimensional sketched detergent bottle and cap assembly.

TIP: *The 2D sketch is intended to represent complex geometry, not duplicate it! Keep it simple. If complex geometry is partially complete or your database contains a similar part, create a drawing containing only the views you want in your layout. Use* **Interface → Export (DXF, IGES...)** *to export the views. Then return to the layout and* **Interface → Import** *the views and proceed with additional sketching.*

In addition, assembly mounting references should be established as part of the first goal. The importance of establishing these references will be discussed later. Creating the references requires steps similar to those taken to sketch the 2D representation. While in the Detail menu, select **Create → Datum** in order to sketch a line representing a datum plane. The 2D datum plane will have a "red" and "yellow" side. You will also be prompted to enter the name of the datum. Create the name of the datum with the understanding that it will be used globally (e.g., *LO_MATE* for a layout mating reference). The side will be important later when referencing components back to the layout for automatic assembly. (The yellow side in the component will match the yellow side in the layout.) An Axis will also be created in the same manner (e.g., *LO_CENTER*). The two assembly mounting references provide the necessary information to place the optional cap. If the component was not circular, an additional reference would be necessary.

The assembly mounting references.

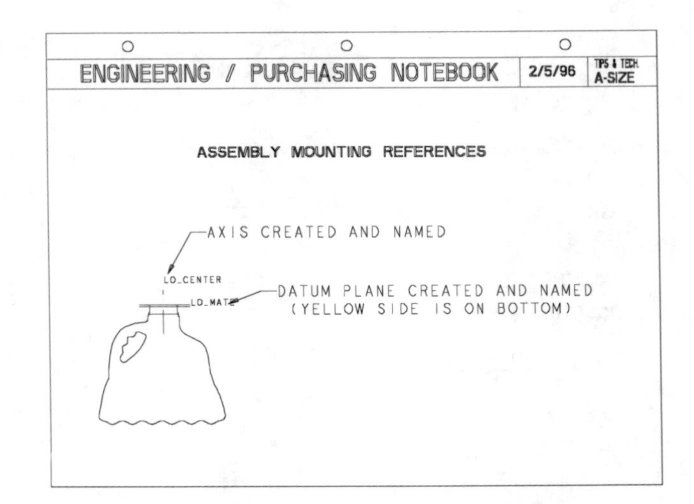

Balloon is another option that can add clarity to the layout. This option is also located in the **Detail → Create** menu. This option allows you to create a balloon and associate a name to the number displayed in the balloon. The names entered for each balloon are then automatically listed.

Automatic balloon listing.

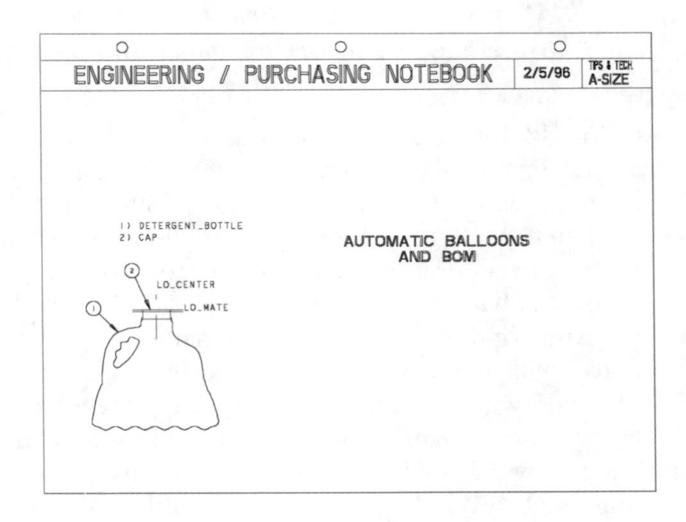

Design Guidelines

The next step is to define the design guidelines for the project. Design guidelines are comprised of a set of notes created to explain important requirements, procedures, or parameters for project development. The guidelines are created by selecting **Detail → Create → Note**. The note can be created with or without leader lines or read in from a file (e.g., a file created from a word processor).

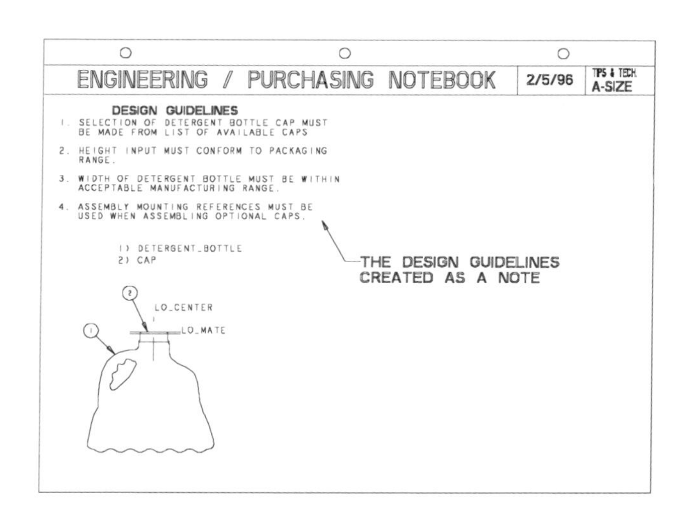

Design guidelines added to the layout.

Charting Available Detergent Caps

At this point, a chart is created to list the available detergent bottle caps. The chart is created using Pro/ENGINEER's table functionality, which is ideal for charts or spreadsheets. From the Layout menu select **Table → Create**. If you use the defaults, a table will be constructed by picking a location on the screen. This screen pick establishes an upper left starting point for the table. Next, select the number of characters for the first column width, and proceed until all column widths have been defined. Select Done (middle mouse button), and then select the number of characters to determine the depth of the first row. Select character numbers for the depth of all rows. When you select Done, the table grid is established. While in the Table menu, select Mod Rows/Cols and establish each column's text justification (e.g., Center | Middle). Finally, select the Enter Text option from the Table menu. This will allow you to select and enter text in any cell (box) within the table (e.g., CAP_1 and AVAILABLE).

Table chart (available detergent bottle caps).

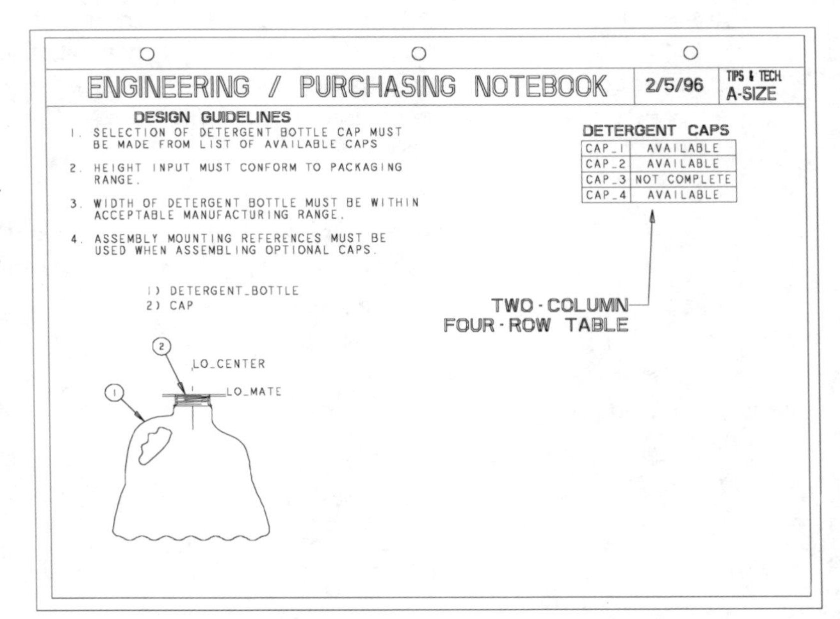

✔ **NOTE:** *Text justification within a column cannot be changed once text has been entered.*

Table Input Section

The table functionality will again be used to create an input section containing critical design parameters. First, table parameters must be established. The three critical parameters for the example layout are cap type, height, and width. To create the parameter for the cap type, access the Layout menu and select **Set Up → Parameter → Create → String** and enter the name for the parameter (e.g., *CAP_TYPE*) and the default value (e.g., *CAP_1*). Once the parameter name has been entered, any string of characters can be input representing the *CAP_TYPE* parameter.

Another technique for creating parameters in a layout will be used to develop the height and width parameters. Both height and width can be created by adding dimensions to the layout. Because the layout is intended to be a graphical notebook, dimensions automatically become parameters in the layout. Select **Detail → Create → Dimension** and pick on the draft entities (2D representation) to create the dimension. During the creation process two prompts will appear. The first asks for the name of the parameter (e.g., *HEIGHT* and *WIDTH*). The second prompts for the default value of the parameter (e.g., 250 and 180, respectively).

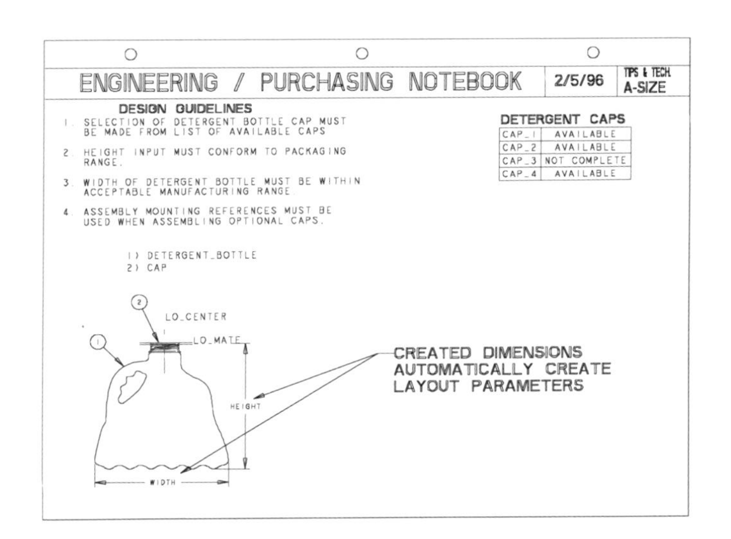

Created dimensional parameters.

Following the same procedures used to create the chart for the available caps, create a table comprised of two columns and three rows. Each row will contain information describing the critical design parameter and the current value for that parameter. In the left column, enter text into each cell that describes what the user needs to input (e.g., cap type, height, and width). In the right column, enter the parameter name preceded by the ampersand (e.g., *&CAP_TYPE*). This technique performs two functions. First, it enables the cell to always reflect the current value of the parameter (e.g., *CAP_1*). Second, it enables the user to simply select within the cell to modify the parameter, thereby providing the "input" value.

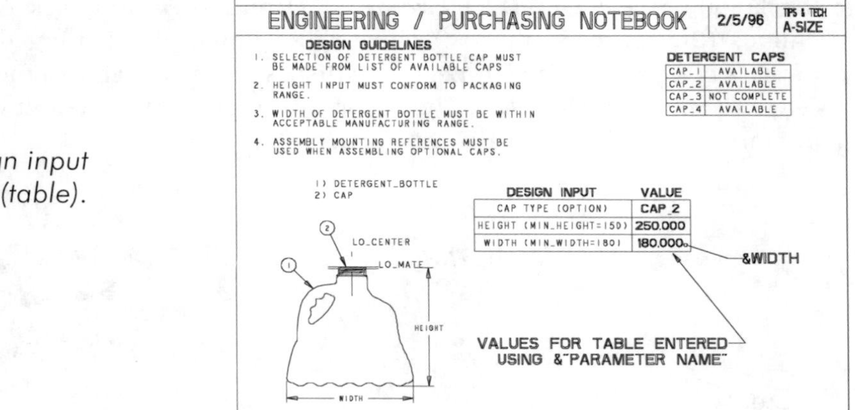

The design input section (table).

Design Checklist

One of the most important reasons for creating a layout is to capture information or parameters for a design. The captured information or parameters can then be used in conjunction with other parameters to provide instant feedback upon regeneration. The feedback can be charted or applied to a spreadsheet in "notebook" form to assist engineering development. A design checklist are created in the layout to describe whether the three critical items meet certain engineering goals.

Three additional "Yes No" parameters will be created. The first will be used to verify whether the *CAP_TYPE* is usable. Select **Set Up** → **Parameter** → **Create** → **Yes No** and enter the name of the parameter (e.g., *USABLE_CAP*). Next, enter the default value for the parameter (i.e., *YES*). Repeat this process for the second parameter, *PKG_RNG* (package range), and the third parameter, *MFG_RNG* (manufacturing range).

Once the parameters have been created for the design checklist, relationships are then added to define the engineering objective. Select **Relations** → **Edit**, and add the following:

```
/*************************************************************
/** DESIGN CHECKLIST **      | = or    & = and
/*************************************************************
IF CAP_TYPE = = "CAP_1" | CAP_TYPE == "CAP_2" | CAP_TYPE
= = "CAP_4"
     USABLE_CAP = YES
     ELSE
       USABLE_CAP = NO
ENDIF
/**
/** HEIGHT (PACKAGE) CHECK **
IF HEIGHT >= 150 & HEIGHT <= 250
   PKG_RNG = YES
   ELSE
     PKG_RNG = NO
ENDIF
/**
/** WIDTH CHECK **
IF WIDTH <= ((((HEIGHT-MIN_HEIGHT)/3) * 2) + MIN_WIDTH)
& WIDTH >= MIN_WIDTH
   MFG_RNG = YES
   ELSE
     MFG_RNG = NO
ENDIF
```

A table is created to describe the checklist parameter, its requirement, and whether or not it is Yes (True) or No (False). In the table, the last column should have text entered with the appropriate parameter called out (e.g., *&USABLE_CAP*). As with the design input "values," the parameter entered with the ampersand (&) will always update upon regeneration and provide automatic feedback to the layout (notebook).

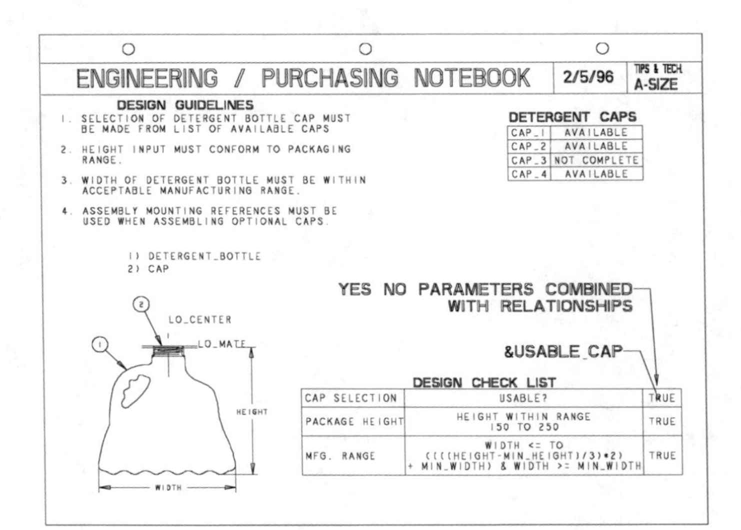

Design checklist.

Cost Evaluation Chart

One additional chart or spreadsheet will be added to the example layout. This spreadsheet will evaluate the cost of the detergent bottle, the chosen cap, and shipping, and provide a total. The spreadsheet is added to the layout to demonstrate how other important information can be applied the notebook.

Similar to the design checklist, four additional parameters are created to set up the cost evaluation chart. Each parameter will be a Number rather than a Yes No. The four parameters are *BOTTLE_COST, CAP_COST, SHIPPING_COST,* and *TOTAL_COST.* During creation of the parameters, any default value can be initially entered (e.g., default value of 35 for *BOTTLE_COST*). The actual value will be determined based on the following relations:

```
/**********************************************************
/** COST EVALUATION RELATIONS **
/**********************************************************
/** CAP COST
IF CAP_TYPE == "CAP_1"
   CAP_COST = 0.35
```

```
    ELSE
    IF CAP_TYPE == "CAP_2"
         CAP_COST = 0.38
         ELSE
         IF CAP_TYPE == "CAP_4"
              CAP_COST =  0.42
              ELSE
                CAP_COST = 0.0
         ENDIF
     ENDIF
ENDIF
/**
/** BOTTLE COST **
BOTTLE_COST = (0.035*(HEIGHT*WIDTH)/1000)
/**
/** SHIPPING COST **
IF HEIGHT >= 150 & HEIGHT <= 200
   SHIPPING_COST = 0.20
   ELSE
   IF HEIGHT > 200 & HEIGHT <= 250
     SHIPPING_COST = 0.30
        ELSE
          SHIPPING_COST = 3.00
   ENDIF
ENDIF
/**
/** TOTAL COST **
TOTAL_COST = CAP_COST + BOTTLE_COST + SHIPPING_COST
```

A table is then created to describe the cost parameter and its current regenerated value. The last column in the table should have text entered with the appropriate parameter called out (e.g., *&BOTTLE_COST*). As shown in the two previous tables, the parameters entered with the ampersand (&) will always update upon regeneration, thereby providing automatic feedback to the layout (notebook).

Cost evaluation spreadsheet.

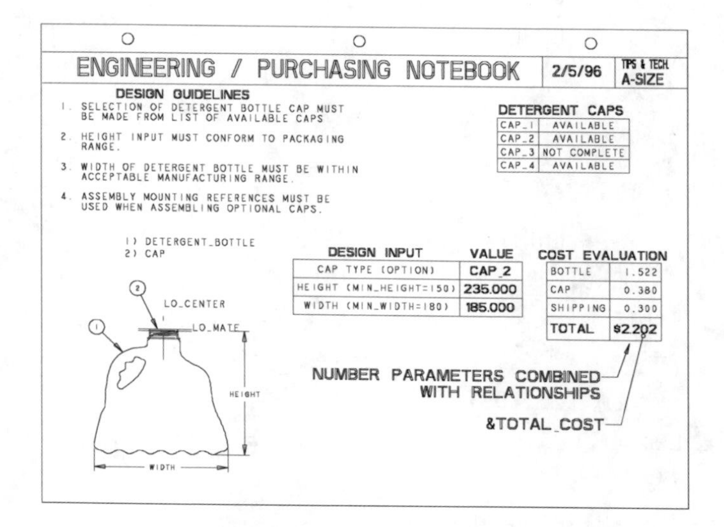

Objectives of Completed Layout

The layout we have created can now provide valuable engineering information such as automatic feedback when critical design inputs are modified. Studies can be performed evaluating both design and cost criteria. Flex or test the layout by entering new values in the design input section of the layout and regenerate.

✗ **TIP:** *A layout can offer performance benefits. Using a layout to establish design criteria and evaluate the "engineering" behind a product's development can save time in the development cycle. Simply put, some engineering aspects for a design can be done in the 2D layout without retrieving a complete complex assembly.*

Layout test 1.

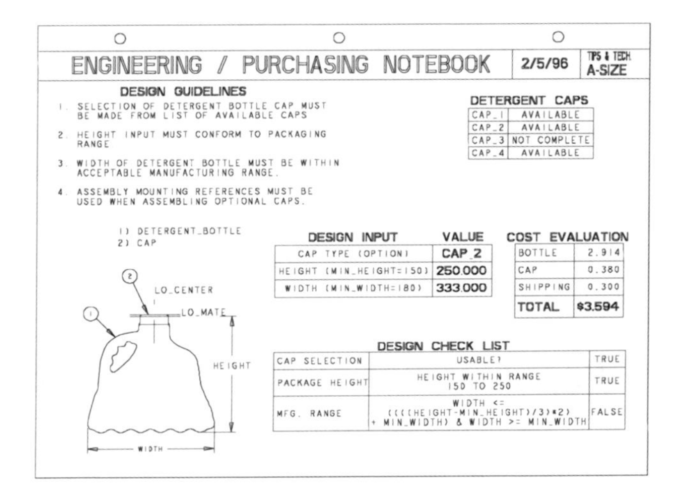

Layout test 2.

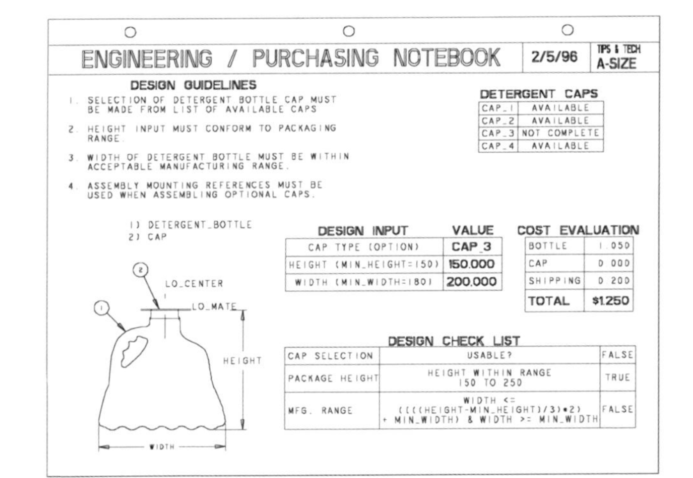

Using Layout Global Information in Components

Another benefit of using Layout is the ability to share the information developed in a layout for components. By using a few functions within a part or assembly, you can reference a layout. These references can then be used to capture the information from the layout. Selected techniques used outside of the layout, and goals for the detergent bottle example, are examined below.

Referencing a Component to the Layout (Declaring)

To obtain information from the layout at the component (part or assembly) level, the component must first be Declared to the Layout. At the part level, within the Part menu, select **Declare → Declare Lay**. The Layout menu will then be displayed listing all available layouts. The reference is established upon selecting the appropriate layout the part is to reference (e.g., *ENG_PURCH_LAYOUT*). At the assembly level, similar steps are taken after selecting Set Up from the Assembly menu. The Declaring of a component to a Layout automatically makes all parameters and assembly mounting references available within the component.

Obtaining Assembly Mounting Information for Automatic Assembly

A major benefit of using Layout is the ability to develop relationships between components and a layout for automatic assembly. This relationship is developed by declaring datums from a part or assembly by name to a declared layout. Once a layout is declared, individual datum features can be associated with individual datum features from the layout. Declaring the same datum in two different components (same reference to a layout datum) creates a placement correspondence between them. There are two ways to declare datums from a component to a layout: using explicit declarations, and creating a table.

In general, explicit declarations are the simplest to create and the easiest to visualize method of declaring component datums to layout datums. We will use explicit declarations for the detergent bottle caps and bottle. We will begin with *CAP_1*, and declare datums within the part to the layout datums. Recall that the layout has two datums defining the assembly mounting references. The first is the *LO_CENTER* describing the axis, and the second is the *LO_MATE* (yellow side down) describing the top of the bottle or inside surface of the caps. In the Part menu, select **Declare → DeclareName** and then select the axis of the *CAP_1* part. At the next prompt, which asks you to enter the name of the global reference, enter *LO_CENTER*. Enter *LO_MATE* at the next prompt for the name of the datum plane. The steps for explicit declarations are then repeated for each additional cap and the bottle.

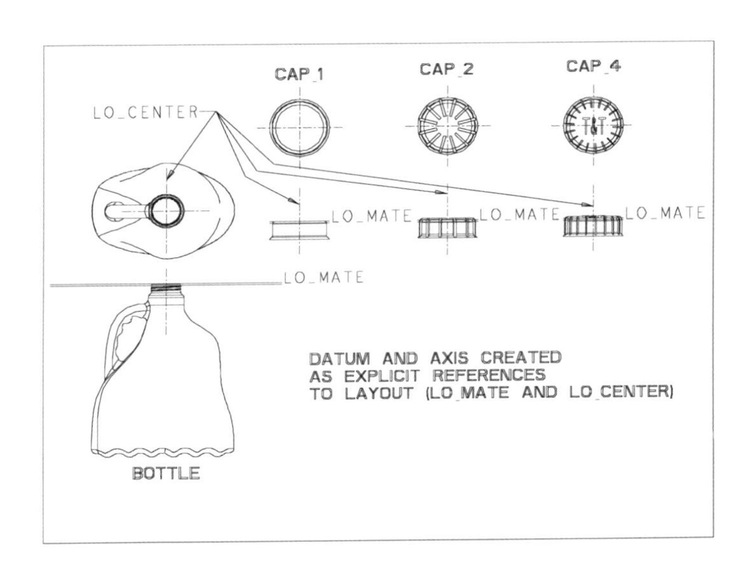

Explicit declarations of component datums to layout datums.

✔ **NOTE:** *When using explicit declarations, you will encounter two limitations. First, you cannot have two datums on the same model with the same explicit declaration (i.e., two datums with the same name). Second, you cannot have one datum with two different declarations (i.e., one datum with two names). If you want to implement either of these practices, you would use the table declarations method. For more information about table declarations, refer to Chapter 7, "2D Layouts and Layout Mode," in the Pro/ENGINEER Assembly Modeling User's Guide.*

Next, create a new assembly and start by assembling the detergent bottle. Assemble the detergent bottle caps one at a time. Each time a new component is assembled, Pro/ENGINEER gathers the correspondences between the new component and the rest of the assembly. If enough correspondences (mounting references) are found, the Auto/Man menu will be displayed, allowing for automatic or manual assembly. Choosing automatic will place the component automatically in the assembly based on its references.

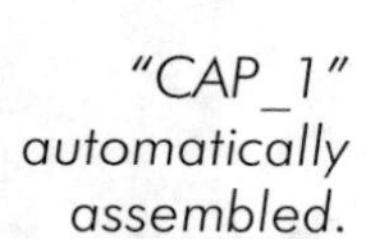

"CAP_1" automatically assembled.

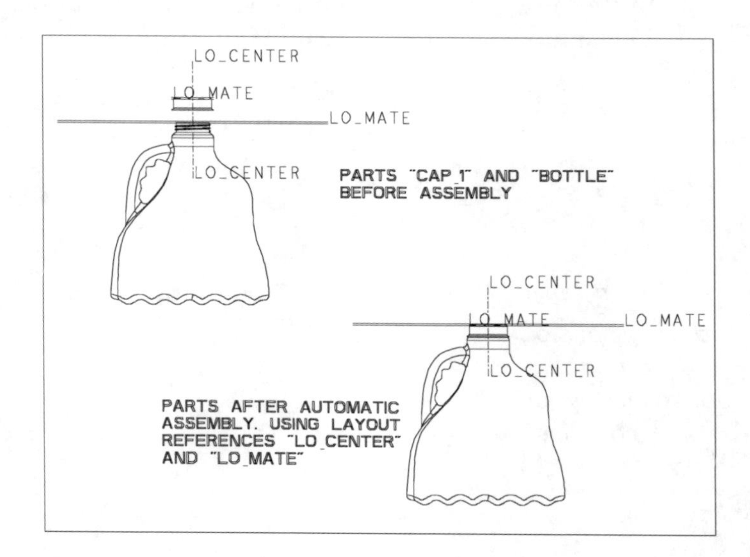

The completed assembly will contain the detergent bottle and three detergent caps, each assembled on top of the other. In the next section, the issue of one cap on top of the other is resolved using information from the layout.

Selecting the Detergent Cap Used in the Final Assembly

To narrow the detergent cap selection to one, the assembly will be declared to the layout (*ENG_PURCH_LAYOUT*). Select **Set Up → Declare**, and choose the appropriate layout. Next, a simple Pro/PROGRAM is created using the reference *CAP_TYPE* from the layout to determine which detergent cap should be assembled. If the user enters a *CAP_TYPE* that is not available, no cap will be assembled. From the Assembly menu, select **Program → Edit**. Add If - Else conditional statements referencing the parameter *CAP_TYPE* to the following information:

```
VERSION XX.0
REVNUM XX
LISTING FOR ASSEMBLY LO_ENG_PURCH_ASM
INPUT
END INPUT
RELATIONS
END RELATIONS

 ADD PART 6405_BTL
 INTERNAL COMPONENT ID 1
 END ADD

        IF CAP_TYPE == "CAP_1"
            ADD PART CAP_1
            INTERNAL COMPONENT ID 2
            PARENTS = 1(#1)
            END ADD

       ELSE
             IF CAP_TYPE == "CAP_2"
                ADD PART CAP_2
                INTERNAL COMPONENT ID 3
                PARENTS = 1(#1)
                END ADD

            ELSE
                IF CAP_TYPE == "CAP_4"
                    ADD PART CAP_4
                    INTERNAL COMPONENT ID 4
                    PARENTS = 1(#1)
                    END ADD
                END IF
            END IF
        END IF

MASSPROP
END MASSPROP
```

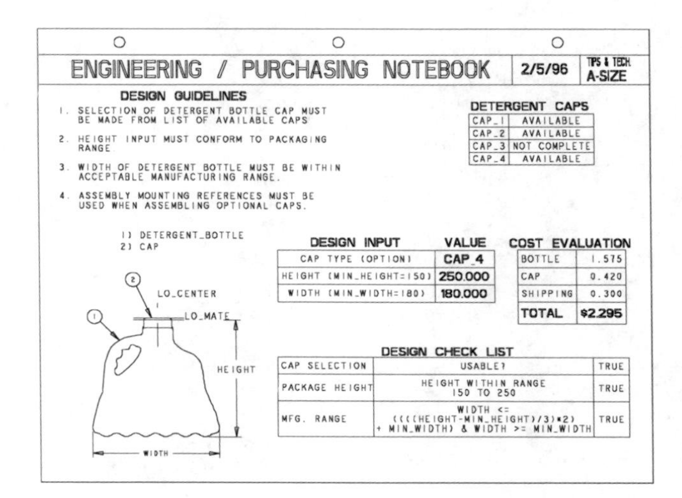

Layout with "CAP_4" entered. (Note the design checklist.)

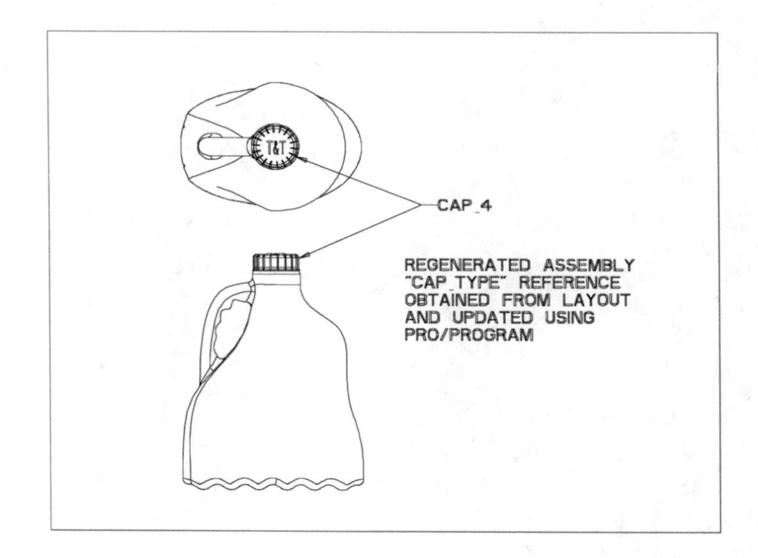

Regenerated assembly referencing "CAP_TYPE" (CAP_4).

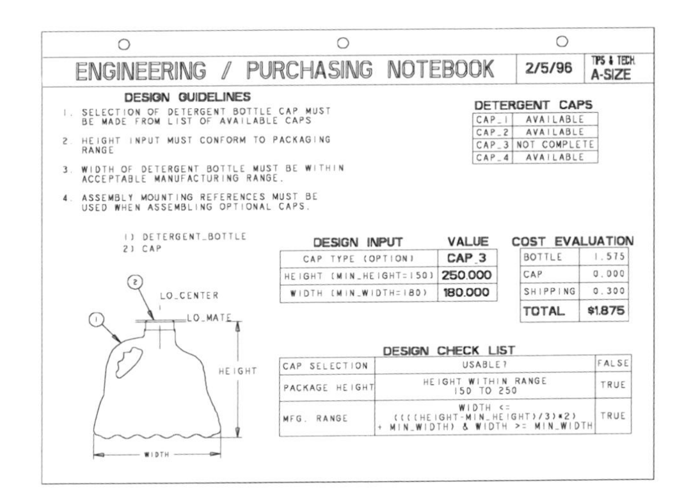

Layout with "CAP_3" entered. (Note the design checklist.)

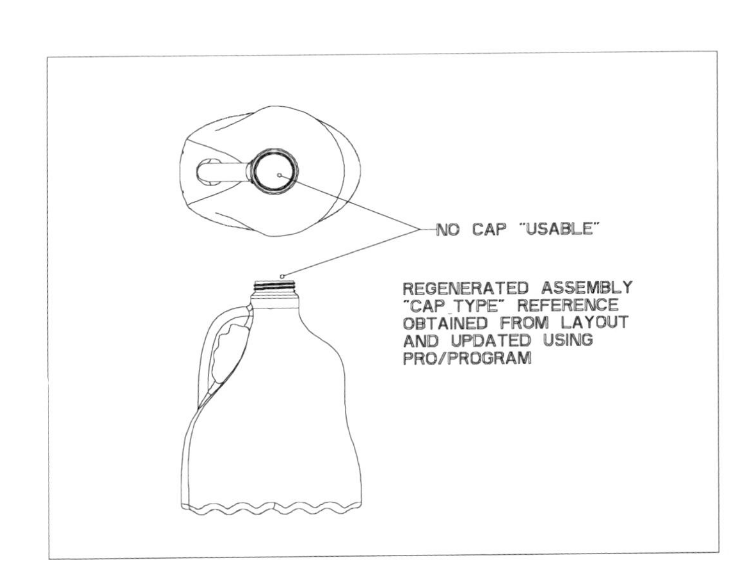

Regenerated assembly referencing "CAP_TYPE" (CAP_3).

Obtaining Critical Parameters for "Height" and "Width"

Two critical dimensions are associated to development of the detergent bottle. The first is a dimension controlling overall height. This dimension has been defined in the layout as "HEIGHT". For packaging reasons the height of the detergent bottle cannot be less than 150, or higher than 250. Values between 150 and 250 are acceptable. For this reason the parameter "HEIGHT" and its value should be passed from the layout to the detergent bottle for regeneration. Second, a similar condition exists for the width of the detergent bottle. For manufacturing reasons described in the layout, the "WIDTH" parameter must satisfy a formula. However, the width can be set to various acceptable values. The value of the parameter "WIDTH" then should be passed from the layout to the detergent bottle for regeneration. From the Part menu, select **Relations → Edit**, and add relations referencing the layout as demonstrated below.

```
/*** Relations for 6405_BTL:
/*****************************************
/** RELATE PART PARAMETERS TO ENG_PURCH_LAYOUT
/*****************************************
/** BOTTLE WIDTH (BTL_WIDTH orginally D9) = LAYOUT WIDTH (WIDTH)
BTL_WIDTH = WIDTH
/** BOTTLE HEIGHT (BTL_HEIGHT orginally D0) = LAYOUT HEIGHT (HEIGHT)
BTL_HEIGHT = HEIGHT
```

After completing the relations, modify the layout "HEIGHT" and "WIDTH" using the design input chart. Return to the detergent bottle (part or assembly) and regenerate.

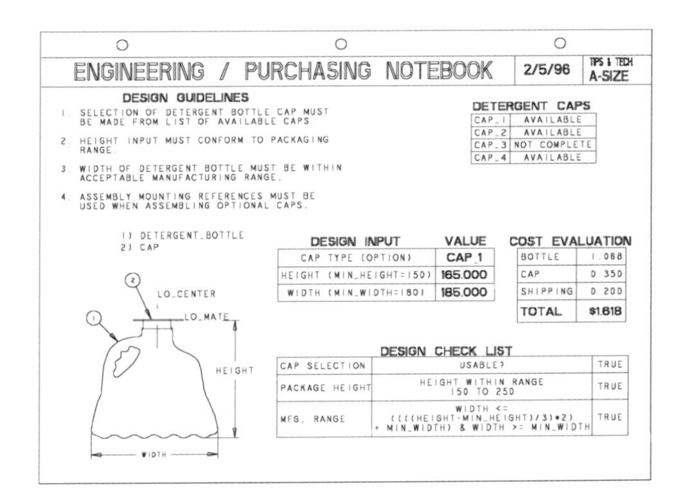

Layout input for HEIGHT and WIDTH.

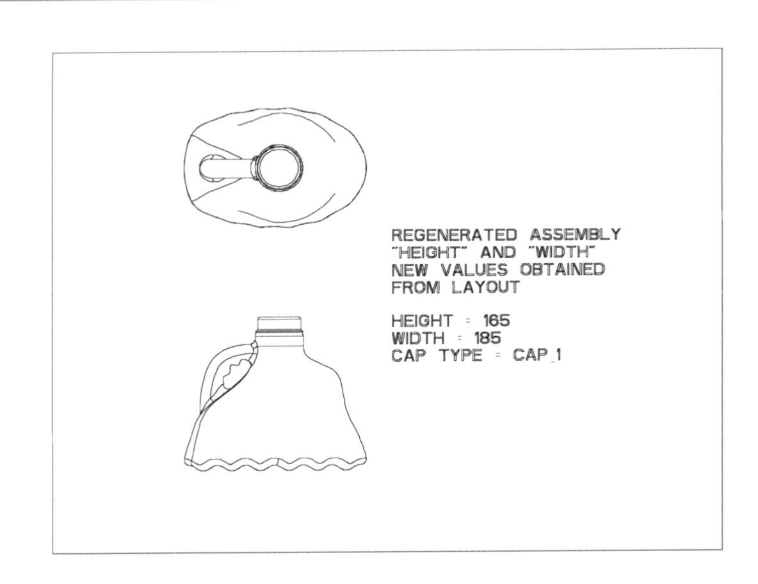

Regenerated assembly.

Layout input for CAP TYPE, HEIGHT, and WIDTH.

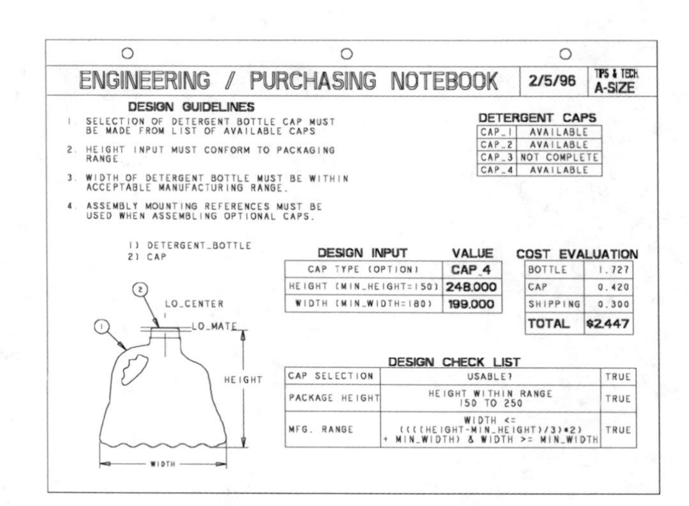

Regenerated assembly.

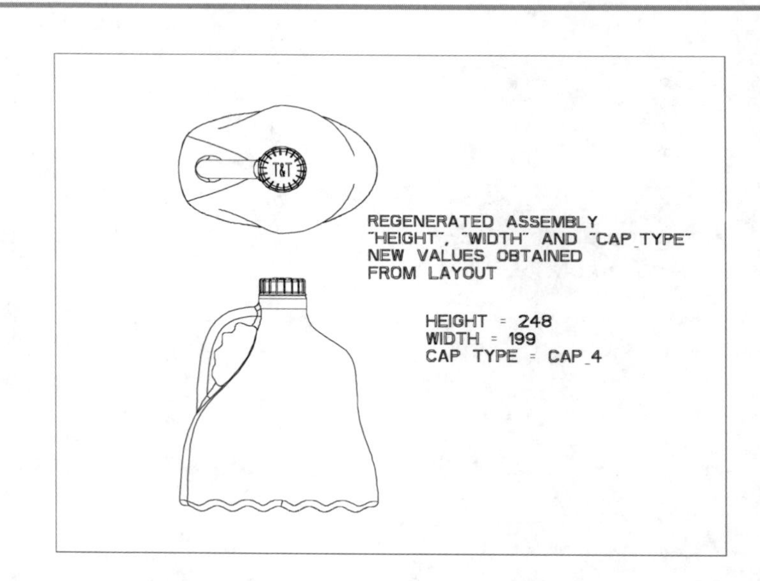

Summary

In this chapter we illustrated the benefits of using Layout mode. We examined how advanced users can dive into Layout mode and come out with an additional tool for improving the engineering process. The example demonstrated how a layout can be used to describe design guidelines, and to perform studies using input parameters, design checks, and even cost evaluations. We also examined how components could be referenced (Declared) to existing layouts to aid in their development. This assistance comes in the form of automatic assembly and relationships linked to the layout's critical design parameters (e.g., HEIGHT). The Layout is an "engineering notebook," and establishes a powerful tool for development from concept to completion.

PART VI Six

Information Tools

Information Tools

Introduction

In this chapter we will discuss various information gathering tools provided by Pro/ENGINEER. The information tools furnish users with methods to identify potential problems and solve them. Two of the simple, but essential, tools for the advanced user are Regeneration Information and Parent/Child. Next, the advanced user typically needs to gather the most information during a feature failure. For information gathering when a feature fails, Resolve Feature Mode (formally called Trim Part) will be examined. By using Resolve Feature Mode, we will illustrate information tools provided during a feature failure and how these tools can be used to resolve any failure.

Regeneration Information (Regen Info)

Regen Info is one of the most important information gathering tools available with Pro/ENGINEER. Regen Info provides you with the ability to "play back" each feature used to create the component. Regen Info is found in the Info menu.

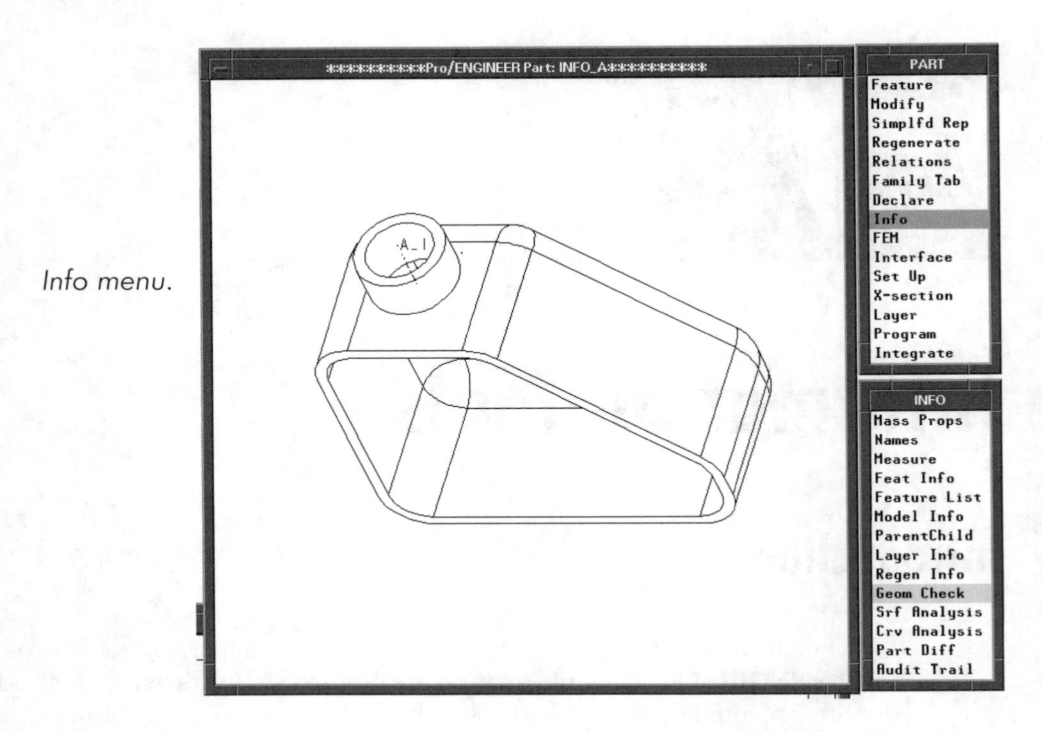

Info menu.

Regen Info's play back ability allows you to step through features and review how each feature was created, thereby obtaining the development sequence or thought process used during creation. During the play back, the user has the ability to select various options including the following:

- ❑ Where to start the "play back" (Beginning or Specify).
- ❑ Information on the current feature (Feature Information window will appear).
- ❑ Showing Dimensions of the current feature.
- ❑ Geometry Check of the current feature (if one occurs).
- ❑ Enter Fix Model before current feature.
- ❑ Skip a group of features during "play back."
- ❑ Continue (step to the next feature, then the next, etc.).

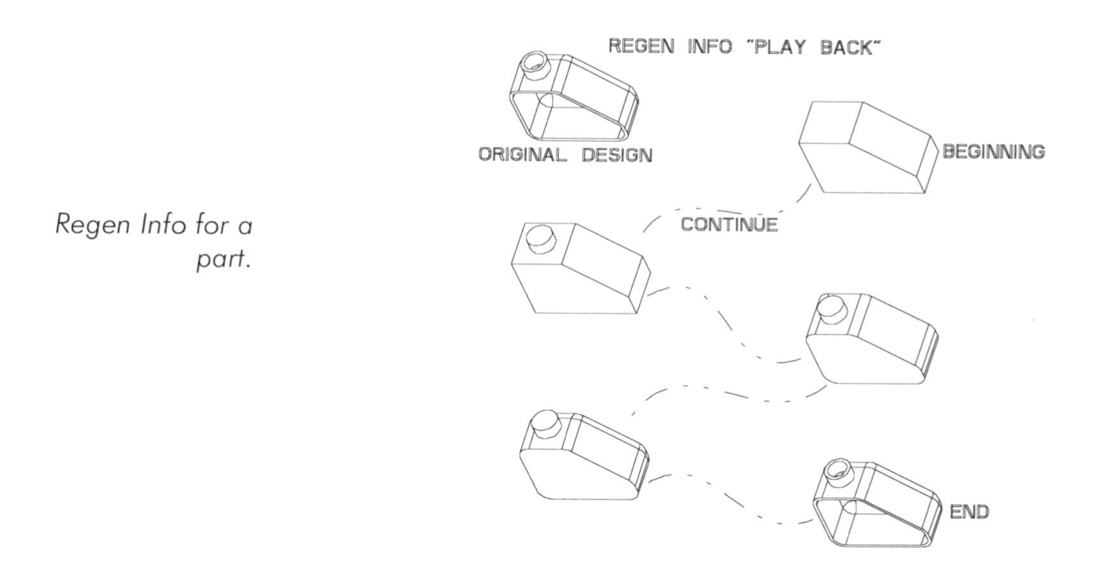

Regen Info for a part.

During Regen Info the option Fix Model is available. This option is extremely powerful, allowing the user to halt the "play back" at any feature and add, remove, or change the model's development and then continue. Fix Model will be discussed in more detail later in the chapter when we examine Resolve Feature Mode.

Parent/Child Information

Parent/Child information, another extremely powerful tool for the advanced user, is used to highlight relationships between features. Before making extensive changes, advanced users will often review a feature's relational information. This simple technique can easily point out potential problems. By selecting **Info → Parent/Child**, the Parent/Child menu will be displayed. Next, you have the option to obtain information on Parents, Children, References, or Child References. By selecting one of the options the information will be displayed in one of the following two methods:

- ❑ File—A text file with feature IDs is written to disk and displayed in the Information Window.

❑ Highlight—The appropriate geometry is highlighted with easy-to-identify color codes, that is, Parents and References (light blue), and Children and Child Ref (magenta for surfaces, edges, or point references, and dark blue for surface mesh).

Parent and Child information.

PARENT/CHILD INFORMATION
2 FEATURE PART
BLOCK AND BOSS

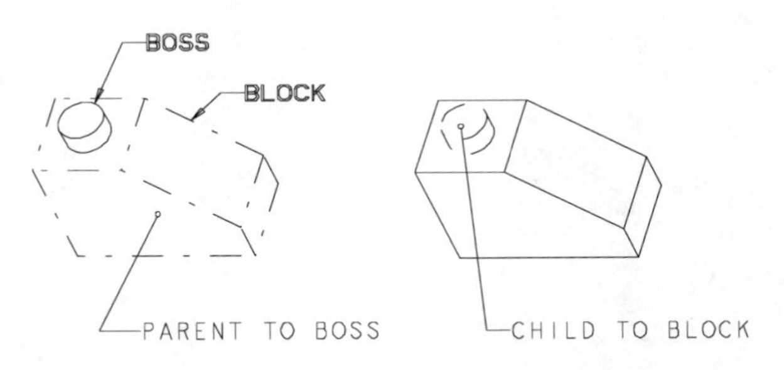

Reference information.

REFERENCE INFORMATION FOR BOSS

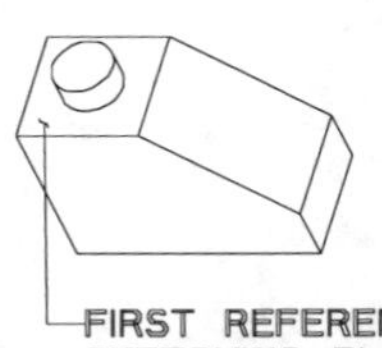

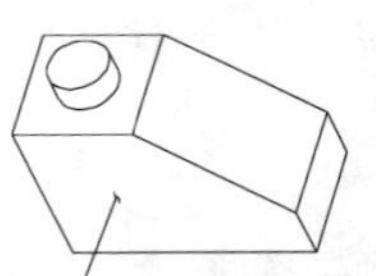

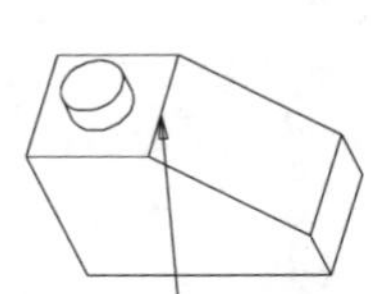

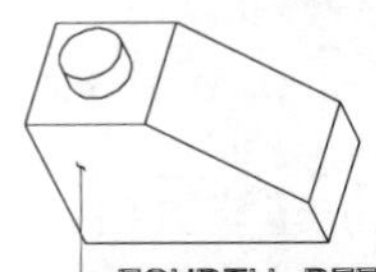

Information via Resolve Feature Mode

Information When a Failure Occurs

The Resolve Feature environment establishes sets of tools to gather information and resolve failures. The causes of failures are usually quite simple, such as in cases where modifications or constructed features conflict with or make other features invalid. The most typical causes of failures appear below:

- Unattached protrusion, and a one-sided edge protrusion is created.
- Resuming features that conflict with another feature or features (e.g., two rounds on the same edge).
- Features intersection becomes invalid because dimensional changes have moved the intersecting surfaces (a Thru Next protrusion modified so that it does not intersect any geometry).
- Relational constraints have been violated.

These typical causes of failures automatically establish a base of information about what happened during a failure.

Working with Resolve Feature Mode

When a failure occurs, you need to know precisely what happened, the cause(s), and how to remedy the situation, among other things. The Resolve Feature environment activates at the instant a failure occurs. The information provided by the Resolve Feature Mode includes the following:

- A basic message displayed in the main message window (e.g., "Could not intersect part with feature").
- The Comprehensive Feature Diagnostic window appears describing the failed feature and its requirements (e.g., attributes, section, direction, etc.).
- Graphical information displaying the model up to its last successful regenerated feature.

An example will help to illustrate the information provided with Resolve Feature Mode. The example part consists of five features.

Part created to illustrate automatic information with Resolve Feature Mode.

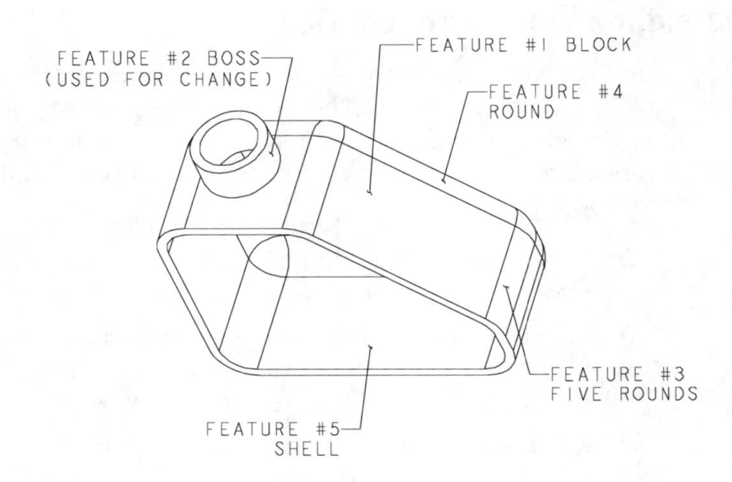

Dimensional changes will be made to both the boss and the shell. The change made specifically to the boss will cause it to fail. Next, the modifications listed above are made and the part is regenerated. A failure occurs demonstrating the automatic information provided.

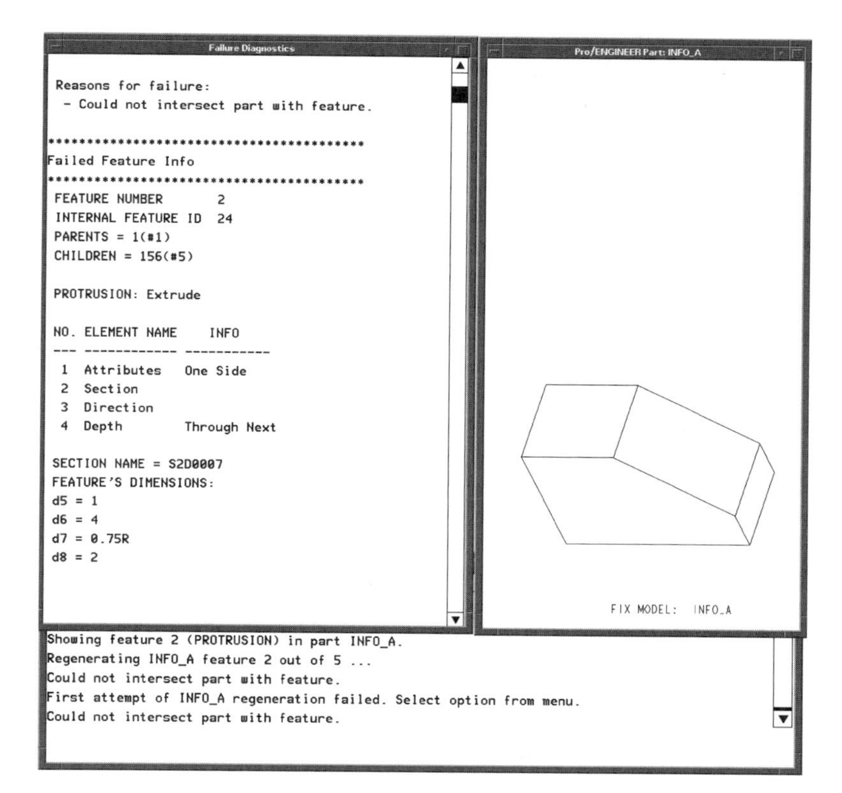

Automatic information.

Additional Information in Resolve Feature Mode

Going beyond Resolve Feature Mode's initial information involves reviewing its other functions. When the Resolve Feature environment is activated due to a failure, the Resolve Feature menu is displayed.

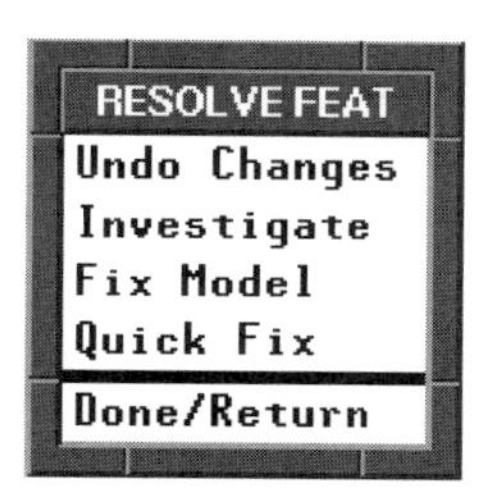

Resolve Feature menu.

In this section we examine the options in the Resolve Feature menu and illustrate the information gathering tools they provide.

The first option is the Undo option. Although Undo does not provide direct information, it does allow you to quickly return to the last successfully regenerated model.

The second option is the Investigate option (sometimes referred to as the Diagnose option). Selecting this option activates additional tools designed for obtaining information about the regeneration failure. These tools can be used based on the Current Model or Backup Model.

- ❑ Current Model—Allows you to perform Investigate options on the current active model (i.e., the failed model).
- ❑ Backup Model—Allows you to perform Investigate options on the backup model. This model will be displayed in a window separate from the current model.

✔ **NOTE:** *When using a Backup Model, you must toggle on the Environment option called RegenBackup. This will cause a copy of the model to be stored to disk before each regeneration (named regen_backup_model####.prt). If the regeneration is successful or Resolve Feature Mode is exited, the backup model is then automatically removed. If the RegenBackup option is not toggled on, the last stored version of the model will be displayed. Individual part size and performance during storing may dictate when to toggle on the RegenBackup option.*

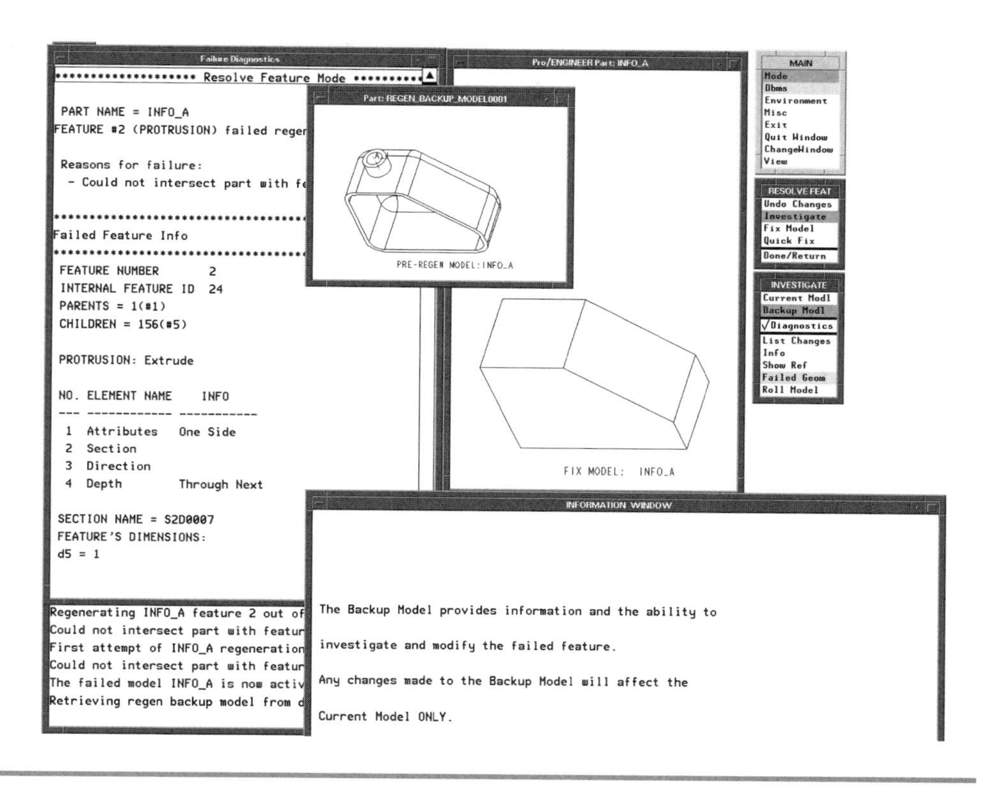

Backup Model.

After choosing either Current Model or Backup Model, the information gathering tools List Changes, Info, Show Ref, Failed Geom, or Roll Model may be used.

- List Changes will show all modified dimensions in a separate window.

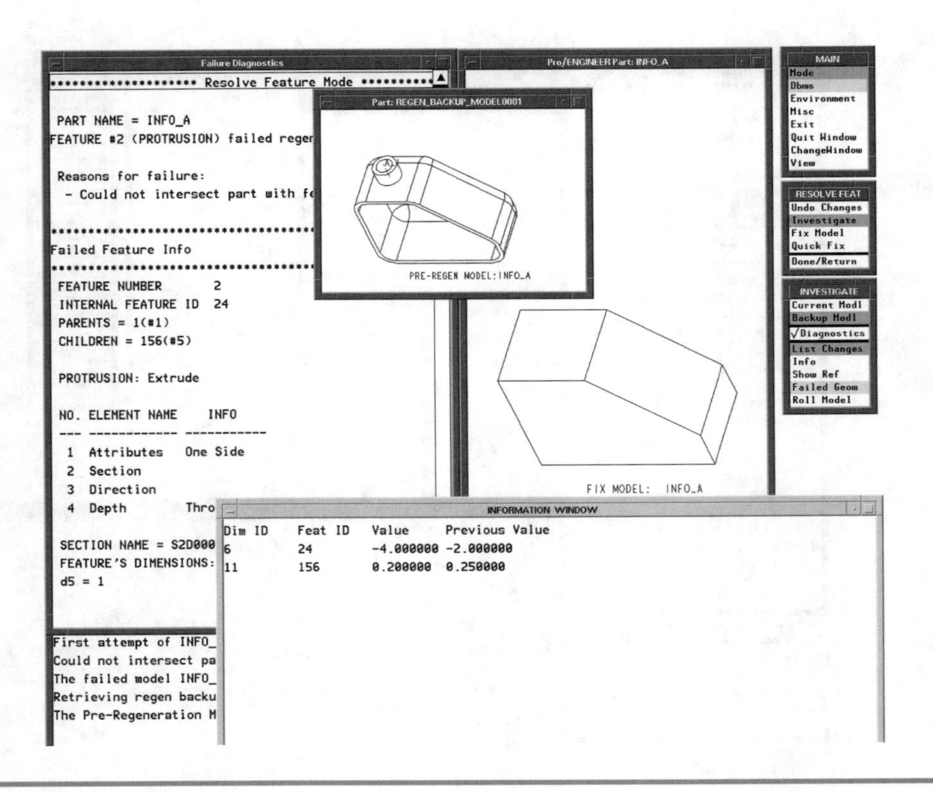

List Changes window (Boss and Shell).

- The Info option provides access to the standard Info menu (shown earlier in the chapter).
- The Show Ref option provides instant access to references of the failed feature. This tool is most beneficial when used with Backup Model (or the last saved version of the part). Show Ref provides a graphical step-by-step review of all references to the failed feature. A good example of this tool's benefits can be demonstrated by creating a round on an edge, and then removing the reference edge the round was originally created on. This information would then be used to help determine what you can do to resolve the failure (e.g., Redefine, Reroute, Suppress, Delete, etc.).

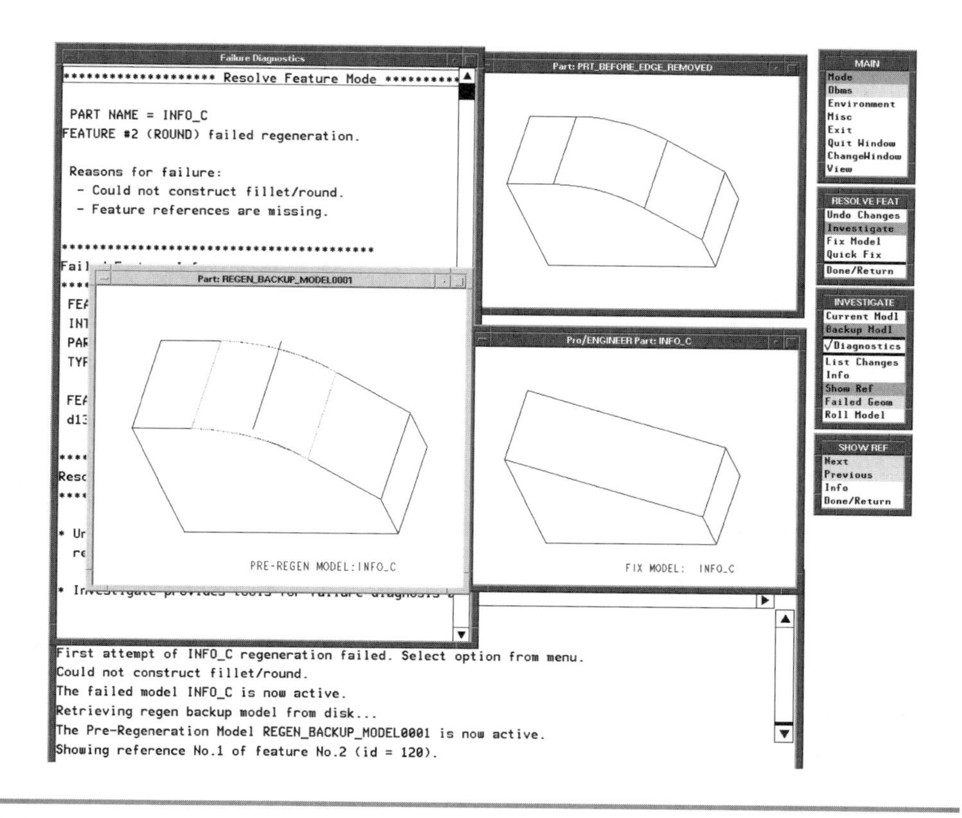

Original model backup showing reference and current failed part (manually copied and displayed for reference only; not part of Resolve Feature Mode).

- ❑ The Failed Geom option provides graphical display of any invalid geometry of the failed feature. Typically the command is unavailable. However, some geometry changes may cause this option to become available (e.g., Shell).

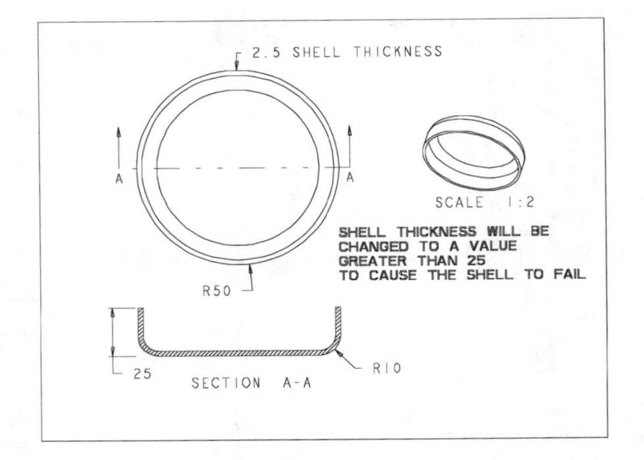

Shell part to be changed.

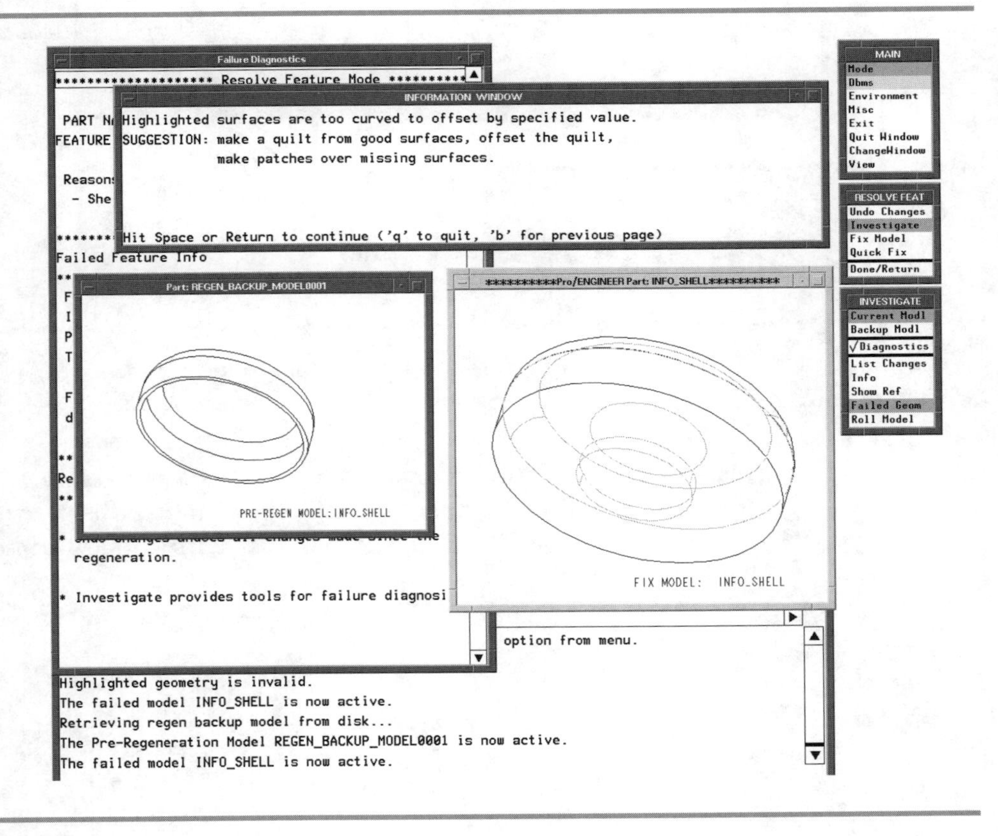

Failed Geom option information.

- The Roll Model option allows you to take the current model to any point in the regeneration of the model. This tool is best used to gather information about the failed feature, features before the failed feature, and features after the failed feature when used with Backup Model (or last saved). Rolling the model to any feature provides information about what happens before, during, and after the failure. This information can help you make decisions about resolving the failure.

The third option in the Resolve Feature environment is the Fix Model option. Fix Model is similar to Investigate in that it offers the ability to work with the Current or Backup model. However, the option differs in that it is more of a geometry tool than an information tool. Exceptions are discussed below.

- Fix Model allows you to access the Feature menu or the Component menu for assemblies. Using options such as Modify, Redefine, and Reroute can provide information. Items such as dimensions and references can be reviewed and changed using this technique.
- Fix Model also provides a Restore option. Restore displays the dimensions, parameters, relations, or other items that have changed in an easy-to-read menu (Sel Mod Dim menu). Any change can then be selected to be restored to previous values. In the case of the shelled part, two dimensions were changed: the radius for the round (10 to 15) and the shell thickness (2.5 to 30).

✗ **TIP:** *The Restore option will work only when using the Current Model. However, you may first select Backup Model to view the display of the backup model, and then proceed by selecting Current Model and Restore. This method is used in the example above.*

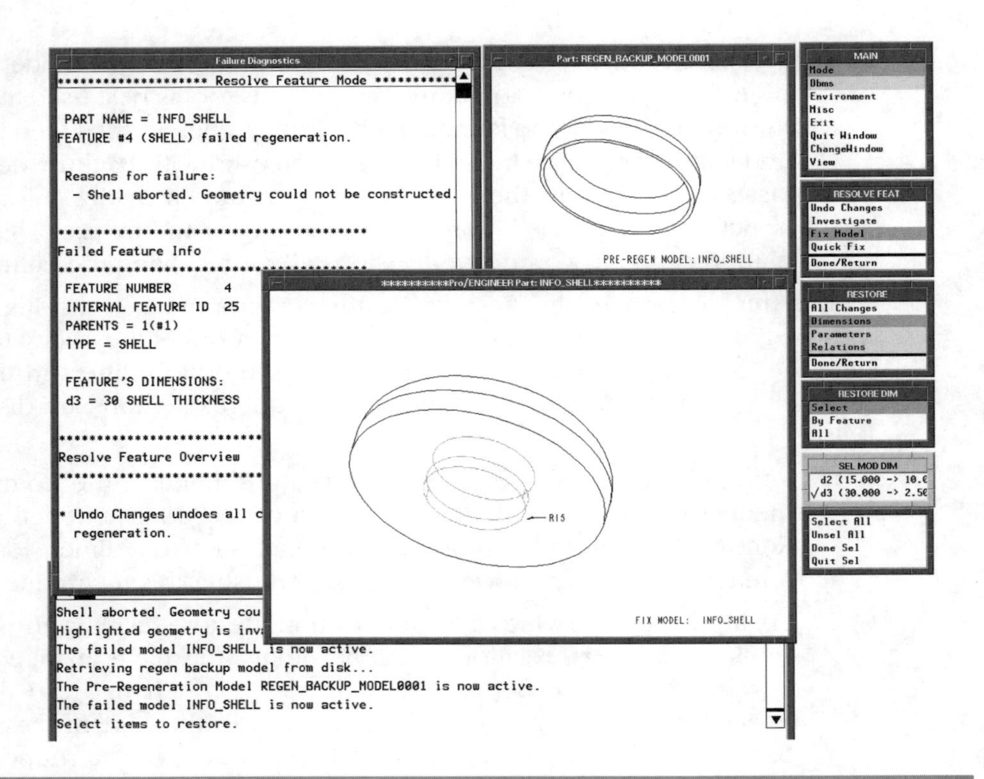

Restore option.

- Fix Model provides menu access to Relations which enables the complete Relations menu. Any information about the relations can be reviewed or changed in the Relations menu.
- The Info menu is also available when using Fix Model. This selection provides access to the standard information menu shown earlier.
- Fix Model also provides complete access to the Set Up, X-Section, and Program menus for obtaining information and making changes.

The fourth and final option available in the Resolve Feature environment is Quick Fix. Quick Fix is similar to Fix Model in that it is more of a geometry tool than an information tool. However, Quick Fix offers Redefine, Reroute, Suppress, and Delete as options to fix the model. Recall that Redefine and Reroute can be used to obtain information about a feature through their ability to show references and dimensions. When working with large assemblies, Quick Fix does not require that the backup model be brought into the session, thereby improving performance while repairing the failure.

✗ **TIP:** *Selecting either Fix Model or Quick Fix without the Environment option RegenBackup toggled on may make the Undo option unavailable. As mentioned earlier, performance should dictate whether to use the Regen Backup option.*

Summary

In this chapter we discussed various information gathering tools provided by Pro/ENGINEER. From the basic tools such as Regen Info and Parent/Child to using Resolve Feature Mode there are an abundance of information gathering techniques for the advanced user. The information tools, especially the tools in the Resolve Feature environment, play an instrumental role in establishing a productive engineering environment ready for change.

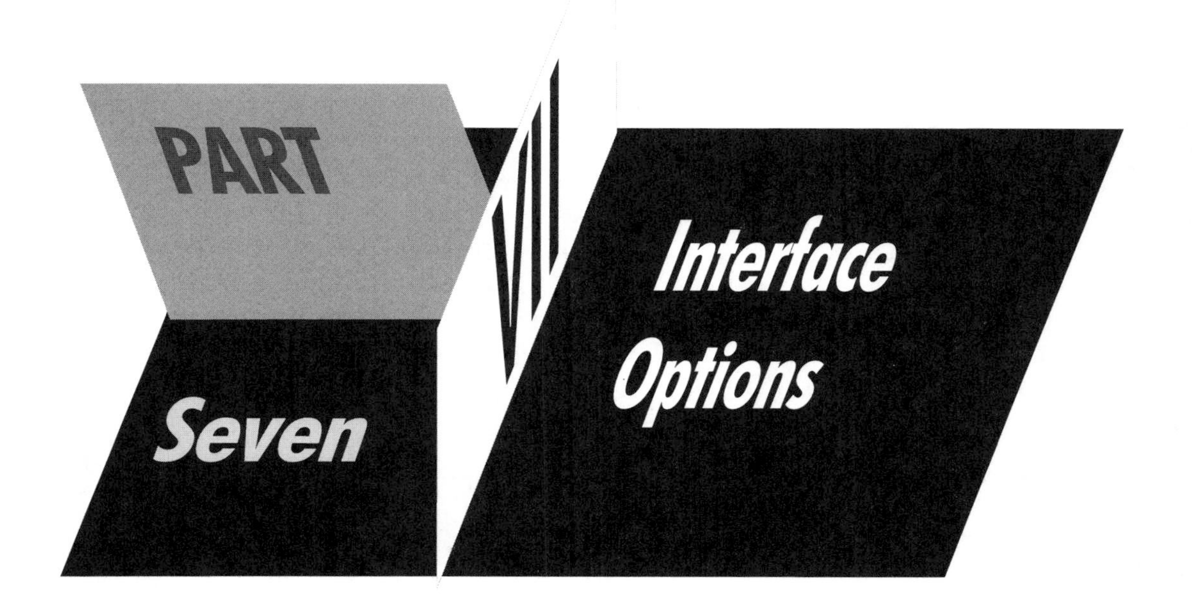
PART
VII
Seven
Interface
Options

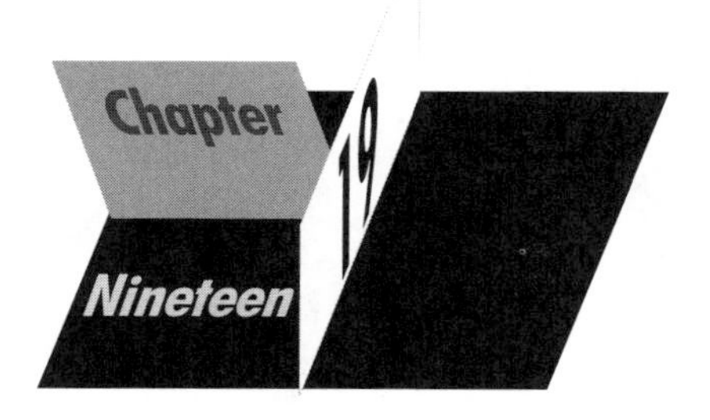

Working with Pro/MESH

Introduction

Finite element analysis is a method used to analytically predict the behavior of an electronic model. Loads and constraints are applied to a finite element mesh model to simulate real life conditions. Results of these "analytical prototypes" are used in conjunction with assumptions, experience, and correlations to test data to assist in engineering decision making. A common thread among all analyses such as structural, thermal, kinematics, and CFD is that they originate with a finite element mesh model (FEM).

Because complicated systems and subsystems are often the focus of analysis in the contemporary workplace, there is a demand for faster and more reliable meshing. Pro/ENGINEER solid models are the perfect vehicle for generating an auto-mesh. Auto-meshing capability is the key to the future for finite element analysis in order to keep pace with the design revolution. This chapter's primary focus is on tips and techniques for improving mesh results within Pro/MESH.

Connecting rod.

Designing for FEM

The most important factor in successfully generating an auto-mesh is the quality of the underlying design model. Whether it be solid or shell elements, 75 percent of most mesh problems originate from a poorly constructed design model. The quality of the model can be further divided into two categories: model clarity and model content. Listed below are some general rules to follow for promoting the best possible auto-mesh.

Model Clarity

In most cases many fine feature details need not be present when meshing the model. These features are typically small rounds and fillets, chamfers, short boundary edges, and small holes. The complexity of the design model should be reduced by de-featuring the model in areas of less concern. Typically, only the areas of focus require fine feature details to be included in the mesh. The rest of the model is used to create elements for mass and stiffness. Different model states are used for design and analysis.

Design state.

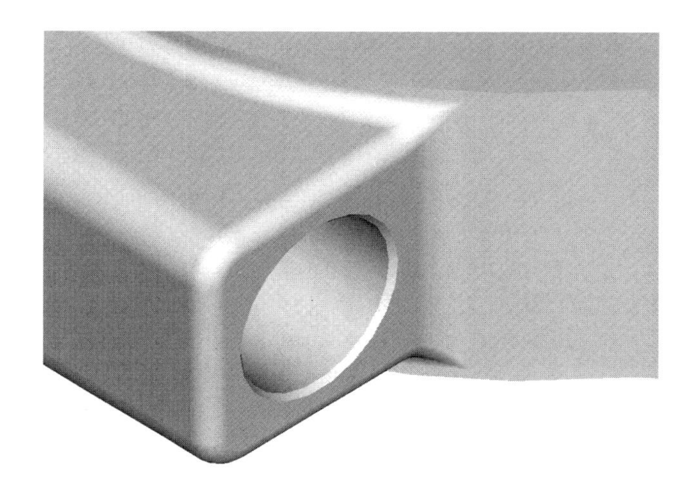

Analysis state.

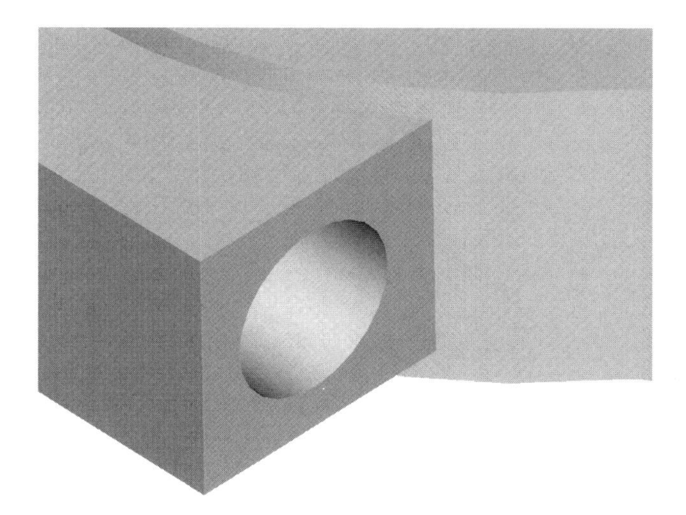

Potential Problems

- Increased mesh times—Excessive detail in the model results in a longer mesh time and increases the risk of failure to generate a mesh. This also results in increased CPU time, RAM use, and swap space.
- Large element counts—Extraordinarily high element counts result in longer solver solution times, thereby reducing analysis efficiency. The meshing generator maps to the smallest feature in the model resulting in high element counts.
- Small mesh elements—A large number of small mesh elements results in an irregular pattern of mesh density. This promotes the creation of a bad transition element from small to large mesh areas, high aspect ratios, and even collapsed elements. Depending on the solver of choice, small mesh elements can account for error messages.

Solutions

- Simplified model—De-feature the model in areas of relatively minor importance. This is accomplished by practicing good design principles such as the following: (1) Placing detail features at the end of the model; (2) Keeping track of unwanted parent/child relationships; (3) Automatically assigning detail features to a layer for quick deletion or suppression; and (4) Creating family table instances of multiple state models.
- Minimize manual pairing of surfaces—When working with thin-walled features, use a global shell command with solid thin features for automatic surface autopairing.
- Geometry checks—Geometry checks indicate that entities (edges, faces) within the model are small relative to the global size of the model. This will also cause the mesh generator to spend a lot of time trying to distinguish the mesh pattern across the small edges. In the process, the mesh generator quite often creates an artificially local high element count or no mesh at all. If possible, eliminate all geometry checks. Return to the model and explicitly align in the sketcher or make reasonable modifications to the geometry.

- Small surface patches (shells)—Minimizing the number of surface patches when surfacing is generally recommended. Extra surface patches result in inconsistent and high mesh densities. The creation of unwanted patches most typically occurs when surfacing from multiple boundary edges. Two ways to reduce or eliminate extra patches follow: (1) Use control points to adjust the flow between hard vertex points when blending between multiple sections. This will cut down on the number of default patches blended between the sections. (Refer to Chapter 11 for more detail on control points.) (2) Create a C2 approximate composite curve through the edges of the desired sections prior to creation of the surface. The outcome is a continuous single-piece, boundary-blended patch which results in better mesh results.

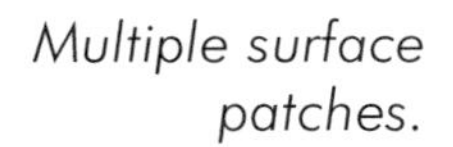
Multiple surface patches.

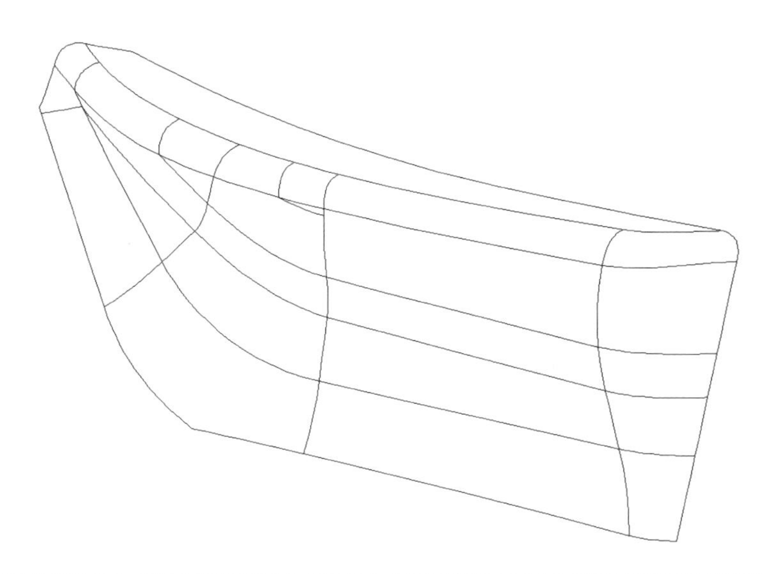

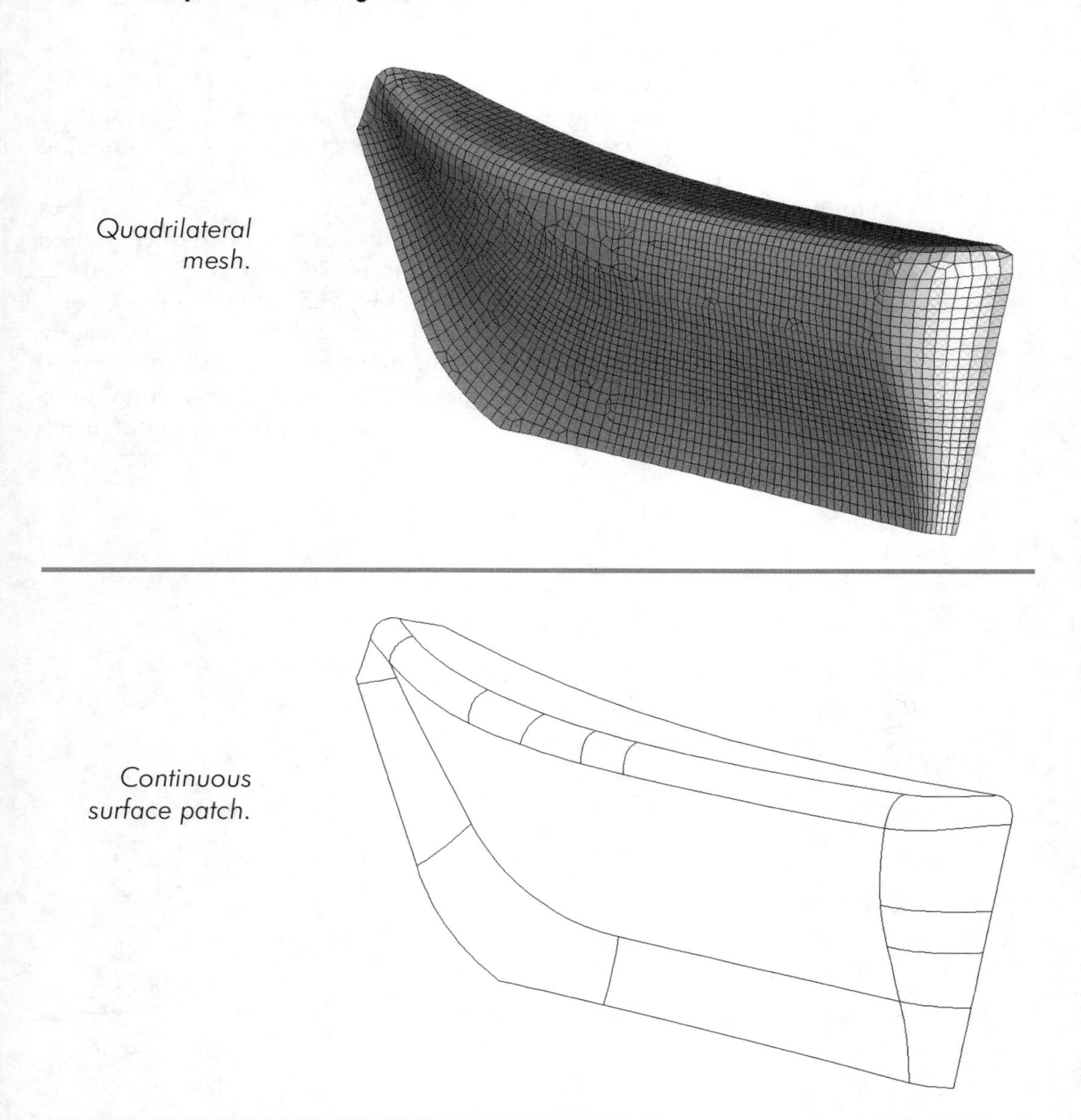

Quadrilateral mesh.

Continuous surface patch.

Model Content

Model content is almost as important as model clarity because the content is the actual geometry used in generating the mesh. Appearing below are suggestions for determining the content of the design model.

- Communicate—Work with everyone involved in the design process (e.g., designer, analyst, test engineer) to determine the required amount of geometry for auto-meshing. Collectively agree on the content and detail of the model that apply to a particular analysis. Preliminary planning may save you a great deal of time and energy trying to mesh unnecessary geometry.
- Develop standards—Develop FEM standards or rules of thumb for the content of the model prior to meshing. For instance, assume that all rounds and fillets smaller than the material thickness on a thin-walled part must be suppressed before shell meshing. This standard will keep everyone informed from the same sheet of paper and promote efficiency by new people in the group.
- Identify appropriate elements—In the preliminary phase of the analysis, identify the elements required for each solid (Tetra4 or Tetra10), shell (triangular or quadrilateral), or beam. Small cylindrical features may be modeled with beams instead of solid meshes, and for mold flow, you may want to work with triangular rather than quadrilateral elements.

Meshing Tips

- Mesh with the Pro/ENGINEER default mesh controls first to determine whether the model will mesh. The objective here is to resolve geometry problems at this point before a tighter mesh control is applied to the model. If you start off with a fine mesh control, it may take 45 minutes to detect a geometry problem compared to ten minutes with the default mesh control.
- Gradually add local mesh control to faces, edges, and hard points before global mesh controls are applied. Because Pro/ENGINEER only provides outside surface local mesh control, this action will help you determine the mesh density before adding any global mesh control.
- Add global minimum and maximum controls to further fine tune the overall mesh of the part. Depending on the part size, introduce a global minimum first, and then a global maximum to help the mesh hone in on an element size average. Pro/ENGINEER takes these inputs into consideration, but may not precisely adhere to the hard number.
- Use a boundary element shell mesh to identify problem areas on the outside surface. Before solid meshing, this practice allows you to fix problem areas which will result in a better solid element mesh.
- Use **Info → Measure → Curve → Short Edge** to identify small edges within the model. This command will highlight on screen all the small edges shorter than a user-defined parameter. Modify the model to remove these edges.
- When the model absolutely will not mesh, begin to section off the model with a series of cuts. Mesh each section until the problem areas are identified. This is a technique used in isolating problem areas within the solid model.
- Verify that the *FEM_MAX_MEMORY* is set high (750 to 1000 megabytes).
- Utilize as much RAM as possible. Avoid swapping to the disk, if possible.
- Use hardware system analytical and graphical checks to monitor the amount of CPU time, memory, and swap space used during mesh generation. The auto-mesher uses 90% of the memory required for the

operation in the first pass. If a memory error occurs during the first pass, adjust the global maximum mesh control accordingly until the first pass threshold is broken. More times than not, a memory occurs in the first pass. Multiple iterations may be attempted to maximize the size of the mesh subject to the hardware limitations of the machine.

Summary

Using Pro/MESH is an iterative process. There are many ways to manipulate the model to adjust the outcome of the mesh. Predicting the behavior of the mesh generator is difficult. Thus, experience plays a large role in generating an acceptable mesh. The presence of a clean design model is essential before automeshing.

Index

!

2D illustration, Layout Mode goals 256–257, 259, 261, 263, 265, 267
2D tangent lines, surface design initiation, blended surface creation 172–174
3D center line, patterning complex sweeps 106–109
3D curves, fully parametric base cap surface creation 184–185
3D splines
 described 140
 formats 140

A

Actual Length, datum points creation 107–109
adding additional curves, non-planar shapes 51–53
ADV GEOMETRY menu
 Sketcher mode, conics 136–137
 Sketcher mode, splines 137–138
advanced curves and surfaces
 3D splines 140
 color rendering analysis 157–158
 conic curvature accelerations 144–145
 conic sections, described 134–137
 curve analysis, Crvture Disp 143
 curve analysis, described 142
 datum curves, wireframe data read as 141–142
 importing point and curve data 140
 MaxDihedral 155
 multiple curves illustrated 148–151
 porcupine smoothness 156–157
 reflect curves 158
 Sketcher mode, conics 136–137
 Sketcher mode, spline constraints and options 139
 Sketcher mode, splines 137–138
 spline curvature accelerations 145–146
 splines, described 137
 straight line and arc 146–148
 surface analysis tools 153, 155
 surfaces, control points 153, 155
 surfaces, creation from boundaries 151, 153–154
align function, Sketcher 11
Align, Part Mode, concurrent engineering 23–27
analytical surface intersection 170, 172
angular drive dimension, kinematics assemblies, skeleton technique 210–211
arcs
 see Datum curves
Arcs-Fillets, feature reduction 75–76
ASCII files, Sketcher mode, spline constraints and options 139
ASCII points, surface design initiation, importing data 162–163
assemblies, large assembly management
 see Large assembly management
assemblies, making assemblies move
 see Kinematics assemblies
assemblies, master modeling
 see Master modeling
assembly mode
 concurrent engineering for setup and assembly, blueprint illustrated 201

concurrent engineering for setup and assembly, dependencies 201–203
concurrent engineering for setup and assembly, described
concurrent engineering for setup and assembly, drawings 201–203
concurrent engineering for setup and assembly, goals 201–203
concurrent engineering for setup and assembly, parent/child relationships 201–203
concurrent engineering for setup and assembly, Part Mode use 201–203
concurrent engineering for setup and assembly, regeneration 201–203
flexing assembly, guidelines 203–204
flexing assembly, managing large assemblies 203–204
associative derivative models
see Master modeling
auto-meshing capabilities, Pro/MESH 299–300
automating assemblies, kinematics assemblies 221–224

B

B-spline control polygon 137
Balloon option, Layout Mode 258
bending spring creation
blueprint 118–121
Datum Graph feature creation 122–123
deformation drive by additional graph 122–123
deformation, relationship to force 126–127
Helical Sweep feature 117–118
HELIX_GRAPH illustration 122–123
movement established by datum curve feature creation 118–121
Protrusion-Sweep addition 126–127
requirements 118–121
three-point spline, additional curves added 118–121
Variable Section Sweep, regeneration of sketch 124–127
Variable Section Sweep, shape control with 122–123
Variable Section Sweep, Sketcher relationship established 124–126
Variable Section Sweep, spine and x-vector 124–126
Variable Section Sweep, surface creation 124–126
Variable Section Sweep, trajectories created 124–126
blend function, variable section sweep compared 46–47
blended curve intersection, surface design initiation 170, 172
bosses, master modeling process 245–246
Boundary Surfaces
creation from boundaries 151, 153–154
feature reduction, blade design 78–80
Bounding box, simplified Representation 231–233
breakdown of design 4–5
breaking down sketch, regeneration failure 17–18
bump creation, patterning complex sweeps, Variable Section Sweep 109–111
By Rule-Distance, Simplified Representation 233–234
By Rule-Expression, Simplified Representation 234
By Rule-Size, Simplified Representation 231–233

C

center line side profile, fully parametric base cap surface creation 182
chamfers, multiple identical pick and place features 82–83
changes, simple features, effect of changes to 46–47
changing datum curves, non-planar shapes 51–53
Circle functions, Sketcher 12–13
circle, conic sections 134–137
Circle-Construction, Sketcher 12–13
clay scan data, surface design initiation, importing data 162–163
color rendering analysis, curvature analysis 157–158
complex sweeps
  see Patterning complex sweeps
components
  Layout Mode global information use, Declaring 268–271, 273, 275
  placement, concurrent engineering for setup and assembly 201–203
concurrent engineering
  Part Mode 23–27
  setup and assembly 200–201
conditional statements
  see Pro/PROGRAM
configpro file standards 8
configsup file 8
conic curvature accelerations 144–145
conic curves, fully parametric base cap surface creation 179–181, 183–185, 187
conic sections
  circle defined 134–137
  described 126–127, 134–137
  ellipse defined 134–137
  hyperbola defined 134–137
  parabola defined 134–137
conic tangency values, non-planar shapes 58, 60–61
conics
  fully parametric base cap surface creation, web development 190–191
  Sketcher mode 136–137
  surface design initiation, blended surface 172–174
control points, surfaces 153, 155
coordinate system, surface design initiation 162–163
cost evaluation chart creation, Layout Mode 264, 266–267
cross section creation, non-planar shapes 55–58
Crvture Disp, curve analysis 143
Curvature Display
  curve analysis 143
curvature, Sketcher mode, spline constraints and options 139
curve analysis
  Crvture Disp 143
  Curvature Display 143
  described 142
curve by equation, evaluate feature 66–67
curve data
  importing 140
curve deviation, surface analysis tools 153, 155
curves
  evaluate feature, curve creation 64–65
  feature reduction, Projected Curve 76–77, 79, 81
  framing, described 165–169
  graph features, adding curvature with 58–59
  Local Group 112–113, 115
  master modeling process 245–246
  multiple curves illustrated 148–151
  non-planar shapes, adding curves 55–58, 63–64
curves, datum curves
  see Datum curves

D

datum curves
  3D splines, importing wireframe data 141–142
  bending spring movement 118–121
  blueprint 46–47
  creation, fully parametric base cap surface creation 179–181, 183, 185, 187
  creation, patterning complex sweeps 105, 107, 109, 111
  fully parametric base cap surface creation, web development 189, 191, 193, 195
  kinematics assemblies, skeleton technique 208–209
  non-planar shapes 51–53
  straight line and arc illustrated 146–148
  surface design initiation, blended surface 172
  wireframe data read as 141–142
Datum features, kinematics assemblies, skeleton technique 208–209
Datum Graph feature creation, bending spring creation 122–123
datum planes
  fully parametric base cap surface creation, web development 189, 191, 193, 195
  surface design initiation, relating local key sections to datum planes 163–165
datum points
  fully parametric base cap surface creation, web development 189, 191, 193, 195
  patterning complex sweeps, three-dimensional center line 107–109
Declaring, Layout Mode global information use 268–271, 273, 275
default coordinate system, curve by equation, evaluate feature 66–67
default datum planes
  creation, fully parametric base cap surface creation 179–181, 183, 185, 187
  master modeling process 245–246
  surface design initiation 162–163
default_start_data directory 7
deformation, bending spring creation 122–123, 126–127
dependencies, concurrent engineering for setup and assembly 201–203
derivative models, master modeling process 245–246
design blueprint
  breakdown of design 4–5
  cut and paste, avoiding 5–6
  features, keeping simple 6–7
  standards, configpro files 8
  standards, developing 7–8
  standards, directory 7–8
  standards, naming conventions 9
design breakdown 4–5
design evaluation
  see Layout Mode
design intent, Layout Mode 253–255
dimensioning
  Part Mode, concurrent engineering 23–27
  Sketcher 11
  Sketcher mode, spline constraints and options 139
directory/configpro file 8
Distance, Simplified Representation, By Rule-Distance 233–234
drafts, multiple identical pick and place features 82–83
drawings, concurrent engineering for setup and assembly 201–203
drive wheel diameter, kinematics assemblies, skeleton technique 208–209

duplicating design
- Mirror 81–82
- Surface-Transform 82–83
- Surface-Transform-Rotate-Copy 81–82
- Translate 81–82

E

Edit Design, Pro/PROGRAM, golf club development 93–95
ellipse, conic sections 134–137
end points, Sketcher mode, spline constraints and options 139
entities
- feature reduction, adding entities 74–76
- Sketcher 17–18

Envelope, Simplified Representation 237–238
evaluate feature, non-planar shapes 64–65
explicit declarations, Layout Mode, global information use in components 268–271, 273, 275
Expression, Simplified Representation, By Rule-Expression 234
Extrude option, non-planar shapes 53–55
extruded surface, three-dimensional center line creation 106–107

F

Failed Geom option, Resolve Feature menu 291–292
failures
- parts failure 27
- regeneration failure, Sketcher 17–18
- Resolve Feature mode 285–287, 289, 291, 293

feature reduction
- blade design, adding Round feature 80
- blade design, Boundary Surfaces 78–80
- blade design, duplicating design intent with Surface-Transform 82–83
- blade design, merging surfaces 78–80
- blade design, multiple identical pick and place features 82–83
- blade design, Projected Curve 76–77, 79, 81
- blade design, protrusion creation 81–82
- blade design, regeneration 82–83
- blade design, revolved surfaces 78–80
- blade design, surface creation 75–76
- blade design, Surface-Merge features 80
- blade design, Surface-Transform-Rotate-Copy, duplicating with 81–82
- breaking down design into features 73
- design breakdown into features 73
- entities, adding 74–76
- need for 71–72
- requirements used to establish blueprint 71–72
- Revolved Protrusion, feature creation 74–76
- Rounds, adding 75–76
- Sketcher entities, adding by portion 74–76
- surface use 71–72
- symmetrical sketch, mirroring half 74–76

features
- cut and paste, avoiding 5–6
- evaluate feature creation 64–65
- multiple identical pick and place features 82–83
- naming conventions 9
- Part Mode, concurrent engineering 23–27
- shell function, grouping features 39–41
- simple features, advantages 6–7
- simple features, effect of changes to 46–47

FEM
- meshing tips 306

model clarity 300–301, 303, 305
model content 305–306
problems 302
solutions 302
fillet curves, fully parametric base cap surface creation 194
finishing features
master modeling process 245–246
master modeling, telephone assembly example 246–249
finite element analysis, Pro/MESH 299–300
Fix Model option 281–283, 293–294
flanges, master modeling process 245–246
flexibility
need for 4–5
non-planar shapes 46–47
patterning complex sweeps, relating design variables 113–115
shell function 29–30
flexing
assembly mode 203–204
master modeling process 245–246
part 23–27
part, non-planar shapes illustration 49–51
patterning complex sweeps, three-dimensional center line 107–109
Pro/PROGRAM, golf club development 93–95
flow chart of Sketcher 18
formats, Layout Mode 253–255
fully parametric base cap surface creation, described 177–178

G

global assembly definition, master modeling 245–246
Graph feature
Local Group 112–113, 115
patterning complex sweeps, Variable Section Sweep 109–111
graph features and relations
adding curvature with 58–59
controlling conic tangency values 58, 60–61
Grid Info, Sketcher 13–14

H

height parameters, Layout Mode 274–276
Helical Sweep feature, bending spring creation 117–118
helical threads, non-planar shapes 68–69

I

IGES curves, surface design initiation, importing data 162–163
IGES data
3D splines, importing wireframe data 140
fully parametric base cap surface creation 177–178
importing data, surface design initiation 162–163
inflection points, surface analysis tools 153, 155
Info option, Resolve Feature menu 290
Info-SrfAnalysis, surface analysis tools 153, 155
information tools
parent/child information 283–284
Regen Info 281–283
Resolve Feature mode, information provided 285–287, 289, 291, 293
Resolve Feature mode, menu 287–289
Resolve Feature mode, uses 285–287, 289, 291, 293
input statements, Pro/PROGRAM controlling a cube 89–90

creation 91–93
insert function, shell function, adding features 41–43
Insert, Part Mode, concurrent engineering 23–27
interior points, Sketcher mode, spline constraints and options 139
Investigate option, Resolve Feature menu 287–289

K

key features, master modeling 245–246
key sections, surface design initiation, breaking data into simple shapes 163–165
kinematics assemblies
  creating the kinematic assembly 212–214
  described 208–209
  generating kinematics in assembly by modifying skeleton part 212–214
  skeleton technique, angular drive dimension 210–211
  skeleton technique, benefits 212–214
  skeleton technique, Datum Curve 208–209
  skeleton technique, Datum features and Surfaces 208–209
  skeleton technique, described 208–209
  skeleton technique, drive wheel diameter 208–209
  skeleton technique, link arm 210–211
  skeleton technique, SLA suspension 214–215
  skeleton technique, slide feature 210–211
  SLA suspension, adding the knuckle 220
  SLA suspension, automating 221–224
  SLA suspension, described 214–215
  SLA suspension, initial skeleton design position 215–216
  SLA suspension, long arm skeleton development 216–217
  SLA suspension, short arm skeleton development 218
  SLA suspension, testing 220
knot sequence, splines 137
Known Dimension tool, Sketcher 14–15

L

large assembly management 203–204
  Layering at part and assembly level 225, 227
  Master Representation use 228
  Retrieve Representation use 230
  Simplified Representation use 227
  Simplified Representation, accessing functions 230
  Simplified Representation, bounding box 231–233
  Simplified Representation, By Rule command to select components 230–231, 233, 235
  Simplified Representation, By Rule-Distance 233–234
  Simplified Representation, By Rule-Expression 234
  Simplified Representation, By Rule-Model Name 230–231, 233, 235
  Simplified Representation, By Rule-Size 231–233
  Simplified Representation, By Rule-Zone 235–236
  Simplified Representation, Envelope creation 237–238
  Simplified Representation, Envelope part as Substitute 239–241
  Simplified Representation, other representations used to create 239–241
  skeleton parts 225, 227

suppressing unneeded components 225, 227
tools for performance enhancement 225, 227
wireframe use 225, 227
Layering
configpro files, adding layers to 8
large assembly management 225, 227
master modeling process 245–246
surface design initiation 163–165
Layout Mode
2D illustration of components 256–257, 259, 261, 263, 265, 267
assembly mounting reference establishment 256–257, 259, 261, 263, 265, 267
Balloon option 258
chart creation 259–260
cost evaluation chart creation 264, 266–267
creating layout 253–255
design checklist creation 260–264
design guideline definition 259
design intent 253–255
global information use, Declaring 268–271, 273, 275
global information use, height and width parameters 274–276
global information use, obtaining assembly mounting information 268–271, 273, 275
global information use, selecting design for final assembly 270, 272–273
goals 256–257, 259, 261, 263, 265, 267
table charts 259–260
table input section 260–264
uses 253–255
left hand rotation, bending spring creation using Variable Section Sweep 124–127
light sources, reflect curves 158
Line functions, Sketcher 12–13
Line-center line, Sketcher 12–13
linear and logarithmic Gaussian color-shaded display 153, 155
link arm, kinematics assemblies, skeleton technique 210–211
List Changes option, Resolve Feature menu 289–290
loadpoint/text/configpro file 8
loadpoint/text/configsup file 8
Local Group option
patterning complex sweeps, bump pattern creation 112–113, 115
patterning, shell function 39–41

M

master blended quilt, fully parametric base cap surface creation 195
master modeling
described 243–244
global definition of assembly 245–246
process description 245–246
process overview 245–246
telephone assembly example 246–249
uses 243–244
Master Representation, large assembly management 228
master web quilt, fully parametric base cap surface creation, web patches 191–192
Mate, Part Mode, concurrent engineering 23–27
mating features, master modeling process 245–246
MaxDihedral, surface analysis tools 155
merge
feature reduction, merging surfaces 78–80
master modeling process 245–246
meshing tips, FEM 306
Mirror, duplicating design 81–82
model management, surface design initiation 163–165

Modify command, Pro/PROGRAM, adding relationships 88
Modify-DimCosmetics-Symbol, Pro/PROGRAM, controlling a cube 87, 89
Move command, Sketcher 14, 16–17
multiple identical pick and place features, feature reduction 82–83
multiple patches, quilting a surface 153, 155

N

names
- naming convention standards 9
- Simplified Representation, By Rule-Model Name 230–231, 233, 235

negative or positive value, shell function 31–32
non-planar shapes
- changing design, adding additional curves 51–53
- changing design, adding curve 55–58
- changing design, breakdown 51–53
- changing design, changing datum curves 51–53
- changing design, cross section creation 55–58
- changing design, Extrude option 53–55
- changing design, parent/child relationships 51–53
- changing design, projected curve creation 53–55
- changing design, projected curves 51–53
- changing design, redefining variable section sweep 55–58
- changing design, two-point spline 55–58
- datum curves and surfaces as blueprint 46–47
- evaluate feature 64–65
- feature reduction, surface creation 75–76
- graph features and relations, adding curvature with 58–59
- graph features and relations, controlling conic tangency values 58, 60–61
- graph features and relations, sweeping cross section 58, 60–61
- handle creation, adding curves 63–64
- illustration, design 47, 49
- illustration, flexing the part 49–51
- illustration, trajectory selection 49–50
- illustration, variable section sweep 47, 49
- rounds addition 67–68
- shell feature addition 68–69
- threads addition 68–69
- variable section sweep 46–47

nose curves, fully parametric base cap surface creation 181, 186–187

O

offset datum planes, surface design initiation, blended surface 170, 172
Offset Edge function, Sketcher 11
Offset, datum points creation 107–109
On Curve-Length Ratio, patterning complex sweeps, datum points creation 107–109
overbuilt curves, framing, described 165–169

P

parabola, conic sections 134–137
parameters
- Evaluate feature, non-planar shapes 64–65
- Layout Mode, cost evaluation chart 264, 266–267

Layout Mode, height and width parameters 274–276
Layout Mode, table input section 260–264
Pro/PROGRAM, adding parameter with Program-Edit Design 96–98
Pro/PROGRAM, creation 91–93
parent/child relationships
concurrent engineering for setup and assembly 201–203
design breakdown 4–5
features, simple 6–7
Info-Parent/Child 283–284
non-planar shape illustration 47, 49
non-planar shapes 51–53, 61–62
Part Mode, concurrent engineering 23–27
Part Mode
concurrent engineering 23–27
concurrent engineering for setup and assembly 201–203
failures 27
flexing part 23–27
sample 23–27
single database structure 23–27
parting geometry, master modeling, telephone assembly example 246–249
parting surfaces, master modeling, telephone assembly example 246–249
parts
creation, shell function 29–30
large assembly management, Layering at part level 225, 227
master modeling process 245–246
Simplified Representation, Envelope creation 237–238
patches
fully parametric base cap surface creation, web patches 190–191
patches, patch structure
see Fully parametric base cap surface creation
patterning complex sweeps
blueprint 104–105
bump creation, regenerating section 109–111
bump creation, relationship to Graph feature creation 109–111
bump creation, Sketcher coordinate system 109–111
bump creation, Variable Section Sweep 109–111
bump pattern creation using Local Group option 112–113, 115
Datum Curve creation 105, 107, 109, 111
design requirements 105, 107, 109, 111
illustration 103–104
relating design variables 113–115
spline creation 105, 107, 109, 111
three-dimensional center line creation with Surface and Projected Curve 106–107
three-dimensional center line creation, extruded surface 106–107
three-dimensional center line, datum points creation with On Curve-Length Ratio 107–109
three-dimensional center line, flexing values 107–109
three-dimensional center line, Split 107–109
three-dimensional center line, splitting into portions 107–109
planar tangent line curve, fully parametric base cap surface creation 182
planes, master modeling process 245–246
point data
importing 140
points
Local Group 112–113, 115

master modeling process 245–246
Sketcher 12–13
polynomial arcs, splines 137
porcupine sectional curvature display, surface analysis tools 153, 155
porcupine smoothness, surface analysis tools 156–157
Pro/Legacy module
importing wireframe data 142
Pro/Legacy module, importing wireframe data 142
Pro/MESH
auto-meshing capabilities 299–300
FEM, model clarity 300–301, 303, 305
FEM, problems 302
FEM, shells 303
FEM, solutions 302
finite element analysis 299–300
Pro/PROGRAM
content areas 85–86
controlling a cube, adding relationships with Modify command 88
controlling a cube, incorporating Pro/PROGRAM with Program-Edit Design 89–90
controlling a cube, input statements 89–90
controlling a cube, modifying symbolic value with Modify-DimCosmetics-Symbol 87, 89
controlling a cube, protrusion sketched as square 87, 89
golf club development, adding conditional statement 91–93
golf club development, adding features 98–102
golf club development, adding parameter with Program-Edit Design 96–98
golf club development, blueprint and design intent 91–93
golf club development, flexing design 93–95
golf club development, input parameter addition for cavity back option 98–102
golf club development, input statement creation 91–93
golf club development, loft or number determination 96–98
golf club development, parameter creation 91–93
kinematics assemblies, automating SLA suspension 221–224
uses and limitations 85–86
Program-Edit Design, Pro/PROGRAM
adding parameters 96–98
controlling a cube 89–90
projected curves
non-planar shapes 51–55
patterning complex sweeps 106–107
prototype testing, shell function 29–30, 35
protrusion creation, feature reduction 81–82
Protrusion-Sweep, bending spring creation 126–127
protrusions, Pro/PROGRAM, controlling a cube 87, 89

Q

Quick Fix option, Resolve Feature menu 294
quilting surfaces 153, 155, 172–174

R

redefining variable section sweep, non-planar shapes 55–58
reference geometry
master modeling, telephone assembly example 246–249

Part Mode, concurrent engineering 23–27
Reference Option, Sketcher 13–14
referencing components, Layout Mode, Declaring 268–271, 273, 275
reflect curves, advanced curves and surfaces 158
Regen Info
  described 281–283
  Fix Model option 281–283
  play-back ability 281–283
regeneration
  bending spring creation 124–127
  concurrent engineering for setup and assembly 201–203
  cross section regeneration, updating values by 61–62
  feature reduction, blade design 82–83
  kinematics assemblies 212–214
  kinematics assemblies, SLA suspension testing 221–224
  patterning complex sweeps, bump creation 109–111
  variable section sweep 46–47
relationships
  Layout Mode, global information use in components 268–271, 273, 275
  patterning complex sweeps, relating design variables 113–115
  Pro/PROGRAM, adding relationships with Modify command 88
Resolve Feature mode
  information provided 285–287, 289, 291, 293
  menu, described 287–289
  menu, Failed Geom option 291–292
  menu, Fix Model option 293–294
  menu, Info option 290
  menu, Investigate option 287–289
  menu, List Changes option 289–290
  menu, Quick Fix option 294
  menu, Roll Model option 293
  menu, Show Ref option 290–291
  menu, Undo option 287–289
  uses 285–287, 289, 291, 293
Retrieve Representation, large assembly management 230
Revolved Protrusion, feature creation 74–76
revolved surfaces
  feature reduction, blade design 78–80
  fully parametric base cap surface creation, ring development 193
ribs, master modeling process 245–246
right hand rotation, bending spring creation using Variable Section Sweep 124–127
Roll Model option, Resolve Feature menu 293
rotation, bending spring creation using Variable Section Sweep 124–127
Rounds
  feature reduction 75–76, 80
  feature reduction, blade design 76–77, 79, 81
  multiple identical pick and place features 82–83
  non-planar shapes 67–68

S

SCAN curves
  importing wireframe data 141
scan data, surface design initiation 163–165
Sec Tools, Sketcher 13–14
shaded color spectrum graphs, color rendering analysis 157–158
sharp edge reduction, shell function 36–38
shell features
  master modeling process 245–246
  non-planar shapes 68–69
  Pro/MESH, FEM 303
shell function

adding features with insert function 41–43
changes after testing 35, 37, 39
patterning rib 39
planning shelled parts, flexibility 29–30
planning shelled parts, negative or positive value 31–32
prototype testing 29–30, 35
rounds 36–38
sample 29–30
sharp edge reduction 36–38
stamped metal parts 32–35
uses 29–30
variable section sweep 35, 37, 39
Show Ref option, Resolve Feature menu 290–291
simple shapes, surface design initiation 163–165
Simplified Representation
accessing functions 230–231, 233, 235
bounding box 231–233
By Rule command to select components 230–231, 233, 235
By Rule-Distance 233–234
By Rule-Expression 234
By Rule-Model Name 230–231, 233, 235
By Rule-Size 231–233
By Rule-Zone 235–236
Envelope creation 237–238
Envelope part as Substitute 239–241
large assembly management generally 227
Size, Simplified Representation, By Rule-Size 231–233
skeleton technique
described 208–209
large assembly management 225, 227
Sketcher
align function 11
bending spring creation, Variable Section Sweep 124–126
conics 136–137
construction tools, Circle functions 12–13
construction tools, Line functions 12–13
construction tools, Points 12–13
dimensioning function 11
entities 17–18, 74–76
flexibility, building into sketch 11
flow chart of Sketcher 18
Geom Tools-Move command 14, 16–17
information tools 13–14
Known Dimension tool 14–15
Offset Edge function 11
regeneration failure 17–18
Sketcher entities 17–18
splines 137–138
tools, shell function 41–43
Use Edge function 11
Sketcher coordinate system, patterning complex sweeps 109–111
slab curves
framing, described 165–169
fully parametric base cap surface creation, 179–181, 183, 185, 187
slab surfaces
blended surface creation 168–173
fully parametric base cap surface creation 179–181, 183, 185, 187
slide feature, kinematics assemblies, skeleton technique 210–211
splines
3D splines 140
B-spline control polygon 137
control polygon, with or without 138
degree of polynomial arc 137
described 137
knot sequence 137
patterning complex sweeps 105, 107, 109, 111
Sketcher mode 137–138
surface design initiation, blended surface 172–174
Splits, Local Group 112–113, 115
standards

configpro files 8
developing 7–8
directory 7–8
naming conventions 9
straight line and arc illustrated 146–148
strengthening ribs, shell function, changes after testing 35, 37, 39
sub-assemblies, concurrent engineering for setup and assembly 201–203
subdirectories, standards directory 7–8
Substitute, Simplified Representation, Envelope part as Substitute 239–241
surface design initiation
blended surface creation between slab surfaces 168–173
blended surface creation, 2D tangent lines 172–174
blended surface creation, analytical surface intersection 170, 172
blended surface creation, blended curve intersection 170, 172
blended surface creation, described 168–173
blended surface creation, point reference geometry 170, 172
blended surface creation, shaded surface intersection 168–173
blended surface creation, tangent line projection 172–174
framing, creating curve as own feature 165–169
framing, described 165–169
framing, multiple framed sections 165–169
framing, section illustrated 165–169
framing, wireframe slab 165–169
setup, breaking data into simple shapes 163–165
setup, default datums and coordinate system 162–163
setup, importing data 162–163
surface design project
described 177–178
fully parametric base cap surface creation, described 177–178
fully parametric base cap surface creation, IGES data import 177–178
fully parametric base cap surface creation, illustration 177–178
fully parametric base cap surface creation, leg development 179–181, 183, 185, 187
fully parametric base cap surface creation, ring development 193
fully parametric base cap surface creation, web development 189, 191, 193, 195
surface patches, Pro/MESH, FEM 303
Surface-Merge features, feature reduction 80
Surface-Transform, duplicating design intent 82–83
Surface-Transform-Rotate-Copy, duplicating design with 81–82
surfaces
boundaries, creation from 151, 153–154
Boundary Surfaces, feature reduction 78–80
control points 153, 155
feature reduction 71–72
feature reduction, blade design 76–77, 79, 81
feature reduction, merging surfaces 78–80
kinematics assemblies, skeleton technique 208–209
master modeling process 245–246
patterning complex sweeps, three-dimensional center line 106–107
surface analysis tools, porcupine smoothness 156–157
sweeping cross section, non-planar shapes 58, 60–61

sweeping curves, master modeling process 245–246
sweeps, patterning
see Patterning complex sweeps
symbolic values, Pro/PROGRAM, Modify-DimCosmetics-Symbol 87, 89
symmetrical sketch, mirroring half 74–76

T

table charts, Layout Mode 259–260
tangency constraints
fully parametric base cap surface creation 181
Sketcher mode, spline constraints and options 139
surfaces, creation from boundaries 151, 153–154
tangent line projection, blended surface creation 172–174
templates, Layout Mode 253–255
temporary assembly, master modeling process 245–246
testing, master modeling process 245–246
thin-walled stamped metal parts, shell function 32–35
threads, non-planar shapes 68–69
three-point spline, bending spring creation 118–121
trajectory selection
creation on the fly 49–51
non-planar shapes illustration 49–50
Translate, duplicating design 81–82
two-point spline, non-planar shapes 55–59

U

Undo option, Resolve Feature menu 287–289
Use Edge function, Sketcher 11

V

Variable Section Sweep
bending spring creation 122–126
bending spring creation, spine and x-vector 124–126
bending spring creation, surface creation 124–126
bending spring creation, trajectories created 124–126
Local Group 112–113, 115
non-planar shape illustration 47, 49
non-planar shapes 46–47
non-planar shapes, redefining sweep 55–58
patterning complex sweeps, bump creation 109–111
shell function 35, 37, 39

W

width parameters, Layout Mode 274–276
wireframe data
3D splines, importing 140
datum curves, read as 141–142
wireframe slab, surface design initiation, framing 168–173
wireframes
large assembly management 225, 227
surface design initiation, importing data 162–163

Z

Zone, Simplified Representation, By Rule-Zone 235–236

More OnWord Press Titles

NOTE: All prices are subject to change.

Computing/Business

Lotus Notes for Web Workgroups
$34.95

Mapping with Microsoft Office
$29.95 Includes Disk

The Tightwad's Guide to Free Email and Other Cool Internet Stuff
$19.95

Geographic Information Systems (GIS)

GIS: A Visual Approach
$39.95

The GIS Book, 4E
$39.95

GIS Online: Information Retrieval, Mapping, and the Internet
$49.95

INSIDE MapInfo Professional
$49.95 Includes CD-ROM

Minding Your Business with MapInfo
$49.95

MapBasic Developer's Guide
$49.95 Includes Disk

Raster Imagery in Geographic Information Systems Includes color inserts
$59.95

INSIDE ArcView GIS, 2E
$44.95 Includes CD-ROM

ArcView GIS Exercise Book, 2E
$49.95 Includes CD-ROM

ArcView GIS/Avenue Developer's Guide, 2E
$49.95 Includes Disk

ArcView GIS/Avenue Programmer's Reference, 2E
$49.95

ArcView GIS /Avenue Scripts: The Disk, 2E
Disk $99.00

ARC/INFO Quick Reference
$24.95

INSIDE ARC/INFO, Revised Edition
$59.95 Includes CD-ROM

Exploring Spatial Analysis in Geographic Information Systems
$49.95

Processing Digital Images in GIS: A Tutorial for ArcView and ARC/INFO
$49.95

Cartographic Design Using ArcView GIS and ARC/INFO: Making Better Maps
$49.95

Focus on GIS Component Software, Featuring ESRI's MapObjects
$49.95

Softdesk

INSIDE Softdesk Architectural
$49.95 Includes Disk

Softdesk Architecture 1 Certified Courseware
$34.95 Includes CD-ROM

Softdesk Architecture 2 Certified Courseware
$34.95 Includes CD-ROM

INSIDE Softdesk Civil
$49.95 Includes Disk

Softdesk Civil 1 Certified Courseware
$34.95 Includes CD-ROM

Softdesk Civil 2 Certified Courseware
$34.95 Includes CD-ROM

MicroStation

INSIDE MicroStation 95, 4E
$39.95 Includes Disk

MicroStation 95 Exercise Book
$39.95 Includes Disk
Optional Instructor's Guide $14.95

MicroStation 95 Quick Reference
$24.95

MicroStation 95 Productivity Book
$49.95

Adventures in MicroStation 3D
$49.95 Includes CD-ROM

MicroStation for AutoCAD Users, 2E
$34.95

MicroStation Exercise Book 5.X
$34.95 Includes Disk
Optional Instructor's Guide $14.95

MicroStation Reference Guide 5.X
$18.95

MicroStation for Civil Engineers
A Design Cookbook
$49.95

101 MDL Commands (5.X and 95)
Executable Disk $101.00
Source Disks (6) $259.95

Pro/ENGINEER and Pro/JR.

Automating Design in Pro/ENGINEER with Pro/PROGRAM
$59.95 Includes CD-ROM

INSIDE Pro/ENGINEER, 3E
$49.95 Includes Disk

Pro/ENGINEER Exercise Book, 2E
$39.95 Includes Disk

Pro/ENGINEER Quick Reference, 2E
$24.95

Thinking Pro/ENGINEER
$49.95

Pro/ENGINEER Tips and Techniques
$59.95

INSIDE Pro/JR.
$49.95

INSIDE Pro/SURFACE: Moving from Solid Modeling to Surface Design
$90.00

FEA Made Easy with Pro/MECHANICA
$90.00

Other CAD

Fallingwater in 3D Studio
$39.95 Includes Disk

INSIDE TriSpectives Technical
$49.95

SunSoft Solaris

SunSoft Solaris 2. for Managers and Administrators*
$34.95

SunSoft Solaris 2. User's Guide*
$29.95 Includes Disk

SunSoft Solaris 2. Quick Reference*
$18.95

*Five Steps to SunSoft Solaris 2.**
$24.95 Includes Disk

SunSoft Solaris 2. for Windows Users*
$24.95

Windows NT

Windows NT for the Technical Professional
$39.95

HP-UX

HP-UX User's Guide
$29.95

Five Steps to HP-UX
$24.95 Includes Disk

OnWord Press Distribution

End Users/User Groups/Corporate Sales

OnWord Press books are available worldwide to end users, user groups, and corporate accounts from local booksellers or from SoftStore Inc. Call toll-free 1-888-SoftStore (1-888-763-8786) or 505-474-5120; fax 505-474-5020; write to SoftStore, Inc., 2530 Camino Entrada, Santa Fe, New Mexico 87505-4835, USA, or e-mail orders@hmp.com. SoftStore, Inc., is a High Mountain Press company.

Wholesale, Including Overseas Distribution

High Mountain Press distributes OnWord Press books internationally. For terms call 1-800-4-ONWORD (1-800-466-9673) or 505-474-5130; fax to 505-474-5030; e-mail to orders@hmp.com; or write to High Mountain Press, 2530 Camino Entrada, Santa Fe, NM 87505-4835, USA.

Comments and Corrections

Your comments can help us make better products. If you find an error, or have a comment or a query for the authors, please write to us at the address below or call us at 1-800-223-6397.

OnWord Press, 2530 Camino Entrada, Santa Fe, NM 87505-4835 USA

On the Internet: http://www.hmp.com